I0502932

Concentrations, Loads, and Yields of Select Constituents from Major Tributaries of the Mississippi and Missouri Rivers in Iowa, Water Years 2004–2008

By Jessica D. Garrett

In cooperation with the Iowa Department of Natural Resources

Scientific Investigations Report 2012–5240

U.S. Department of the Interior
U.S. Geological Survey

U.S. Department of the Interior
KEN SALAZAR, Secretary

U.S. Geological Survey
Marcia K. McNutt, Director

U.S. Geological Survey, Reston, Virginia: 2012

For more information on the USGS—the Federal source for science about the Earth, its natural and living resources, natural hazards, and the environment, visit http://www.usgs.gov or call 1–888–ASK–USGS.

For an overview of USGS information products, including maps, imagery, and publications, visit http://www.usgs.gov/pubprod

To order this and other USGS information products, visit http://store.usgs.gov

Suggested citation:
Garrett, J.D., 2012, Concentrations, loads, and yields of select constituents from major tributaries of the Mississippi and Missouri Rivers in Iowa, water years 2004–2008: U.S. Geological Survey Scientific Investigations Report 2012–5240, 61 p.

Contents

Figures

Tables

Conversion Factors

Inch/Pound to SI

Multiply	By	To obtain
Length		
inch (in.)	2.54	centimeter (cm)
inch (in.)	25.4	millimeter (mm)
foot (ft)	0.3048	meter (m)
mile (mi)	1.609	kilometer (km)
Area		
acre	0.004047	square kilometer (km^2)
square mile (mi^2)	2.590	square kilometer (km^2)
Flow rate		
cubic foot per second (ft^3/s)	0.02832	cubic meter per second (m^3/s)
Mass		
ton per day (ton/d)	0.9072	metric ton per day
ton per day (ton/d)	0.9072	megagram per day (Mg/d)
ton per day per square mile [(ton/d)/mi^2]	0.3503	megagram per day per square kilometer [(Mg/d)/km^2]
ton per year (ton/yr)	0.9072	megagram per year (Mg/yr)
ton per year (ton/yr)	0.9072	metric ton per year
Pressure		
inch of mercury at 32°F (in Hg)	0.1333	kilopascal (kPa)
Application rate		
pounds per acre per year [(lb/acre)/yr]	1.121	kilograms per hectare per year [(kg/ha)/yr]

Temperature in degrees Celsius (°C) may be converted to degrees Fahrenheit (°F) as follows:

°F=(1.8×°C)+32

Temperature in degrees Fahrenheit (°F) may be converted to degrees Celsius (°C) as follows:

°C=(°F-32)/1.8

Horizontal coordinate information is referenced to the North American Datum of 1927 (NAD 27).

Specific conductance is given in microsiemens per centimeter at 25 degrees Celsius (µS/cm at 25°C).

Concentrations of chemical constituents in water are given either in milligrams per liter (mg/L) or micrograms per liter (µg/L).

Concentrations, Loads, and Yields of Select Constituents from Major Tributaries of the Mississippi and Missouri Rivers in Iowa, Water Years 2004–2008

By Jessica D. Garrett

Abstract

Excess nutrients, suspended-sediment loads, and the presence of pesticides in Iowa rivers can have deleterious effects on water quality in State streams, downstream major rivers, and the Gulf of Mexico. Fertilizer and pesticides are used to support crop growth on Iowa's highly productive agricultural landscape and for household and commercial lawns and gardens. Water quality was characterized near the mouths of 10 major Iowa tributaries to the Mississippi and Missouri Rivers from March 2004 through September 2008. Stream loads were calculated for select ions, nutrients, and sediment using approximately monthly samples, and samples from storm and snowmelt events.

Water-quality samples collected using standard streamflow-integrated protocols were analyzed for major ions, nutrients, carbon, pesticides, and suspended sediment. Statistical data summaries of sample data used parametric and nonparametric techniques to address potential bias related to censored data and multiple levels of censoring of data below analytical detection limits. Constituent stream loads were computed using standard pre-defined models in S-LOADEST that include streamflow and time terms plus additional terms for streamflow variability and streamflow anomalies. Streamflow variability terms describe the difference in streamflow from recent average conditions, whereas streamflow anomaly terms account for deviations from average conditions from long- to short-term sequentially. Streamflow variability or anomaly terms were included in 44 of 80 site/constituent individual models, demonstrating the usefulness of these terms in increasing accuracy of the load estimates.

Constituent concentrations in Iowa streams exhibit streamflow, seasonal, and spatial patterns related to the landform and climate gradients across the studied basins. The streamflow-concentration relation indicated dilution for ions such as chloride and sulfate. Other constituent concentrations, such as dissolved organic carbon and suspended sediment, increased with streamflow. Nitrogen concentrations (total nitrogen and nitrate plus nitrite) increased with low and moderate streamflows, but decreased with high streamflows.

Seasonal patterns observed in constituent concentrations were affected by streamflow, algae blooms, and pesticide application. The various landform regions produced different water-quality responses across the study basins; for example, total phosphorus, suspended sediment, and turbidity were greatest from the steep, loess-dominated southwestern Iowa basins.

Nutrient concentrations, though not regulated for drinking water at the study sites, were high compared to drinking-water limits and criteria for protection of aquatic life proposed for other Midwestern states (Iowa criteria for aquatic life have not been proposed). Nitrate plus nitrite concentrations exceeded the drinking-water limit [10 milligrams per liter (mg/L)] in 11 percent of all samples at the 10 sites, and exceeded Minnesota's proposed aquatic life criteria (4.9 mg/L) in 68 percent of samples. The Wisconsin standard for total phosphorus (0.1 mg/L) was exceeded in 92 percent of samples. Ammonia standards, current during sample collection and at publication of this report, for protection of aquatic life were met for all samples, but draft criteria proposed in 2009 to protect more sensitive species like mussels, were exceeded at three sites.

Loads and yields also differed among sites and years. The Big Sioux, Little Sioux, and Des Moines Rivers produced the greatest sulfate yields. Mississippi River tributaries had greater chloride yields than Missouri River tributaries. The Big Sioux River also had the lowest silica yields and total nitrogen and nitrate yields, whereas nitrogen yields were greater in the northeastern rivers. The Boyer and Nishnabotna River total phosphorus yields were the greatest in the study. The Boyer River orthophosphate yields were greatest except in 2008, when the Maquoketa River produced the greatest yield. Rivers in southwestern Iowa's Western Loess Hills and Steeply Rolling Loess Prairie ecoregions had the greatest suspended-sediment yields, whereas the smallest yields were in the Big Sioux and Wapsipinicon Rivers. In the 10 Iowa rivers studied, combined annual total nitrogen stream transport ranged from 3.68 to 9.95 tons per square mile per year, and total phosphorus transport ranged from 0.138 to 0.570 tons per square mile per year. Six-month loads relative to fertilizer use ranged from 8 to 56 percent for nitrogen, and 1.0 to 11.1 percent for phosphorus. The smallest loads relative to fertilizer use for both nitrogen and phosphorus occurred in July-December of dry years, and the largest nitrogen and phosphorus loads relative to use were in wet years from January-June.

Introduction

Midwestern agricultural watersheds have been high-lighted as major contributors of nutrients to the Gulf of Mexico, contributing to annual algal blooms and subsequent hypoxia zones in the Gulf (Alexander and others, 2008; Vitousek and others, 1997). Iowa is one of the most productive agricultural areas of the Nation, particularly for corn, soybeans, and swine (U.S. Department of Agriculture, 2009). Fertilizers and pesticides are used to support the substantial crop yields attained in the State. Urban and residential fertilizer and pesticides also are used in Iowa cities and towns, which are generally small with dwellings surrounded by lawns and gardens. Livestock wastes also are applied to crops as an important source of nutrients. Iowa has the largest inventory and sales for swine of any state in the Nation (U.S. Department of Agriculture, 2009), with much of this production in confined animal feeding operations (CAFOs).

The U.S. Geological Survey (USGS) began a project in spring 2004 in cooperation with the Iowa Department of Natural Resources (IDNR) to assess water quality of major rivers in Iowa and to allow estimation of select constituent loads. Other research to estimate loads in Midwest streams has focused on delivery to the Gulf of Mexico, but that research suggests that for many constituents like nitrogen, in-stream processing is minimal for large rivers (Aulenbach and others, 2007; Alexander and others, 2000); thus, transport from the State's major rivers provides a good indication of potential effects to downstream areas. Prior nationwide or Mississippi River sampling networks used to estimate loads have included limited areas of the State, or loads were estimated for unmonitored sites from landscape attributes. For example, The SPAtially Referenced Regression On Watershed attributes (SPARROW) model provides valuable information on nutrient loads as relative subbasin contributions delivered to the Gulf of Mexico (Alexander and others, 2008). SPARROW, however, is based on long-term average conditions, and is not intended to describe temporal patterns or the effects of ongoing and changing land-use practices.

Purpose and Scope

Water-quality data summarized in this report were collected at 10 major rivers draining Iowa from March 2004 through September 2008. The purpose of this report is two-fold: (1) summarize nitrogen, phosphorus, organic carbon, suspended sediment, select ions, and select pesticides data in major Iowa streams; and (2) present estimated loads and yields for select ions, nutrients, and suspended sediment being transported to the Mississippi and Missouri Rivers from major streams in Iowa.

Results presented in this report on stream nutrient delivery can be used to evaluate water-quality conditions during a specific time and to identify emerging water-quality trends. This knowledge will contribute to continuing research into the effects of landscape, land use, and climate on local water quality and downstream delivery of nutrients affecting downstream rivers and the Gulf of Mexico.

Study Area Description

The major basins of Iowa drain into the Mississippi and Missouri Rivers bordering the east and west respective sides of the State, with 6 of the 10 basins studied flowing directly into the Mississippi River. The basins total 50,562 square miles (mi^2) and range in size from 871 to 14,038 mi^2 (table 1, fig. 1). These basins include 75.0 percent of Iowa's total land area, and 17.1 percent of the study basin area extends beyond Iowa into eastern South Dakota and southern Minnesota (U.S. Geological Survey, 2009). Annual average precipitation varies from 22 inches at the northwest extent of the study area to 38 inches toward southeast Iowa (National Oceanic and Atmospheric Administration, 2003). Mean annual streamflows (1979–2008) near the mouths of the 10 basins range from 461 to 10,293 cubic feet per second (ft^3/s) (table 1; U.S. Geological Survey, 2009).

Land use in the study basins is predominately agriculture, and cities and towns are generally small with low population densities. Widespread row-crop agriculture in Iowa accounts for 73 percent of the State, with 86 percent of crop production in corn and soybeans (U.S. Department of Agriculture, 2009, fig. 1). Median and mean 2008 estimated population densities in Iowa cities and towns were 740 and 841 people per square mile (people/mi^2), respectively. County level densities ranged from 9.5 to 746 people/mi^2, averaging 54 people/mi^2 statewide (U.S. Census Bureau, Population Estimates Program, *http://www.census.gov/popest/estimates.php*, accessed January 27, 2011).

The landforms of the study basins (fig. 2) typify glacial plains and karst landscapes, including alluvial valleys, glacial outwash plains, loess hills, and recent and well-weathered glacial areas. Most of the study basins are in the Western Corn Belt Plains ecoregion, defined by U.S. Environmental Protection Agency and characterized by fertile, moist, glacial, commonly calcareous soils formerly covered with tallgrass prairie, now one of the most productive areas for corn and soybeans in the world (U.S. Environmental Protection Agency, 2009b). Within the Western Corn Belt Plains, the Western Loess Hills and the Steeply Rolling Loess Prairies are noteworthy for highly erodible soils and steep slopes. Basins also include areas of the Northern Glaciated Plains, Driftless Area, Central Irregular Plains, and Missouri and Upper Mississippi Alluvial Plains ecoregions. The parts of the Northern Glaciated Plains in the study basins are flat to gently rolling, subhumid grasslands and wetlands with fertile soil, but with climatic limitations on agriculture. The Driftless Area of northeast

Table 1. Site information and streamflow summary for selected major Iowa rivers, water years 2004–2008.

[Water year from October 1 to September 30; ID, identifier; mi², square miles; ft³/s, cubic feet per second; --, not available]

Map ID (fig. 2)	Station number	Station name	Contributing area (mi²)	Mean annual streamflow (ft³/s)					
				1979–2008	2004	2005	2006	2007	2008
1	06485500	Big Sioux River at Akron, Iowa	6,996	2,023	1,209	1,435	2,395	2,189	2,183
2	06607500	Little Sioux River at Turin, Iowa	3,526	1,895	1,338	1,666	1,618	2,184	2,714
3	06609500	Boyer River at Logan, Iowa	871	461	320	266	138	691	941
4	06810000	Nishnabotna River above Hamburg, Iowa	2,806	1,740	1,355	1,047	539	2,665	3,617
5	05490500	Des Moines River at Keosauqua, Iowa	14,038	9,111	7,500	6,447	5,002	12,490	18,680
6	05474000	Skunk River at Augusta, Iowa	4,312	3,175	2,491	2,002	1,046	4,425	7,009
7	05465500	Iowa River at Wapello, Iowa	12,500	10,293	9,386	7,211	5,970	12,700	21,740
8	05422000	Wapsipinicon River near De Witt, Iowa	2,336	2,079	2,288	1,055	1,276	2,584	4,503
9	05418600	Maquoketa River near Spragueville, Iowa	1,632	--	--	--	--	--	--
	05418500	[1]Maquoketa River at Maquoketa, Iowa	1,553	1,251	1,366	673	680	1,254	3,191
10	05412500	Turkey River at Garber, Iowa	1,545	1,276	1,534	733	874	1,562	2,788

[1]Samples for Maquoketa River collected downstream of the steamflow gage.

Iowa, distinct from the corn belt regions, is characterized by limestone and dolomite karst, steep hills, and exposed bedrock bluffs, with little glacial deposits only on ridge tops. Dairy farming is common, with crops grown in patches where slopes permit and pastureland and dense woodlands on steeper terrain. The Central Irregular Plains in southern Iowa are less uniform in topography and land use than the Western Corn Belt Plains to the north; this southern Iowa region includes broader forested riparian areas and areas of previous coal mining. Missouri and Upper Mississippi Alluvial Plains include recent alluvial deposits, as well as broader plains of glacial outwash.

Methods

This section describes protocols and methods for collection and analysis of water-quality data. Continuous streamflow data and results of individual sample measurements and chemical analyses were published annually in the USGS water-data reports (Nalley and others, 2005a, b; U.S. Geological Survey, 2007–2009) and also are available on the USGS National

Water Information System website (NWISWeb) at *http://waterdata.usgs.gov/nwis* (U.S. Geological Survey, 2009).

Water-Quality Data Collection

Water-quality samples were collected and field properties were measured approximately monthly at 10 sites, which were at or near previously established streamflow-gaging stations, near the mouths of basins representing large parts of the State of Iowa. Samples were collected during all seasons beginning March 2004 with an emphasis on storm and snowmelt events to support load and yield estimation. Samples were collected using isokinetic, streamflow-weighted sampling techniques (equal width increment, EWI), except where low streamflow velocities (less than 1.5 ft³/s) or safety considerations (for example, flooding or ice) necessitated adjustments to these protocols to obtain a representative sample of the stream. Maquoketa River samples (map ID 9, station number 05418600) routinely were collected downstream of the streamflow-gaging station at Maquoketa (station number 05418500), due to bridge safety considerations.

Protocols and equipment used for sample preparation, collection, processing, and quality assurance are described

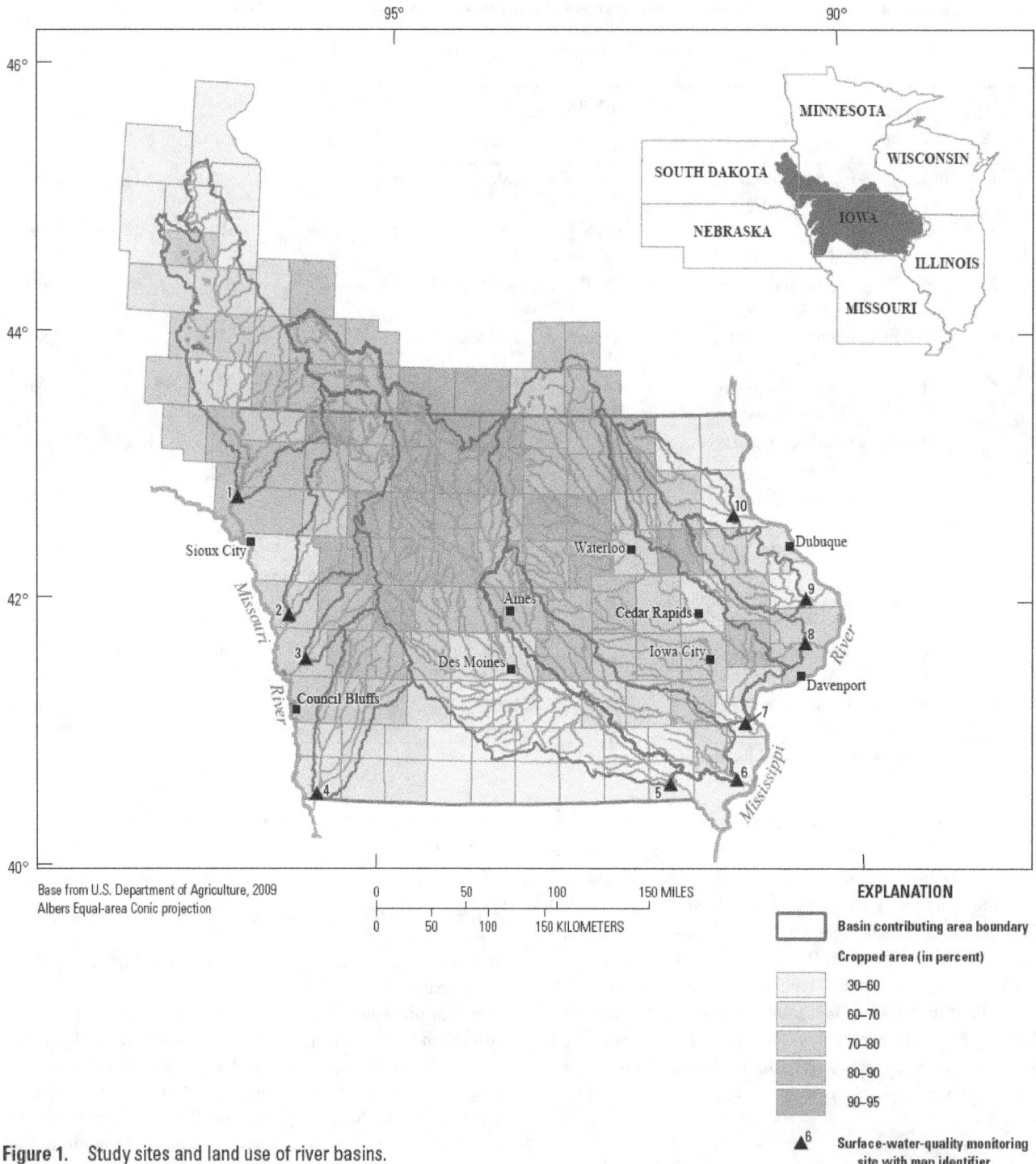

Figure 1. Study sites and land use of river basins.

in the USGS National field manual for the collection of water-quality data (U.S. Geological Survey, variously dated). Methanol, as used to remove residual organic compounds such as pesticides, was omitted from cleaning protocols to avoid erroneously high dissolved organic carbon (DOC) results. Blank, replicate, and spike quality control (QC) samples were routinely collected and analyzed for bias, variability, and analytical recovery. For every 11 environmental samples

collected, 1 QC sample was collected, totaling 16 blanks, 28 replicates, and 9 pesticide spikes. The ratio for replicate sample QC was 20:1. With the omission of methanol, quality-control field blank samples indicated reduced occurrence of DOC without carryover of pesticides.

Samples were analyzed by the USGS National Water-Quality Laboratory in Denver, Colorado and the USGS Sediment Laboratory in Iowa City, Iowa. Field properties,

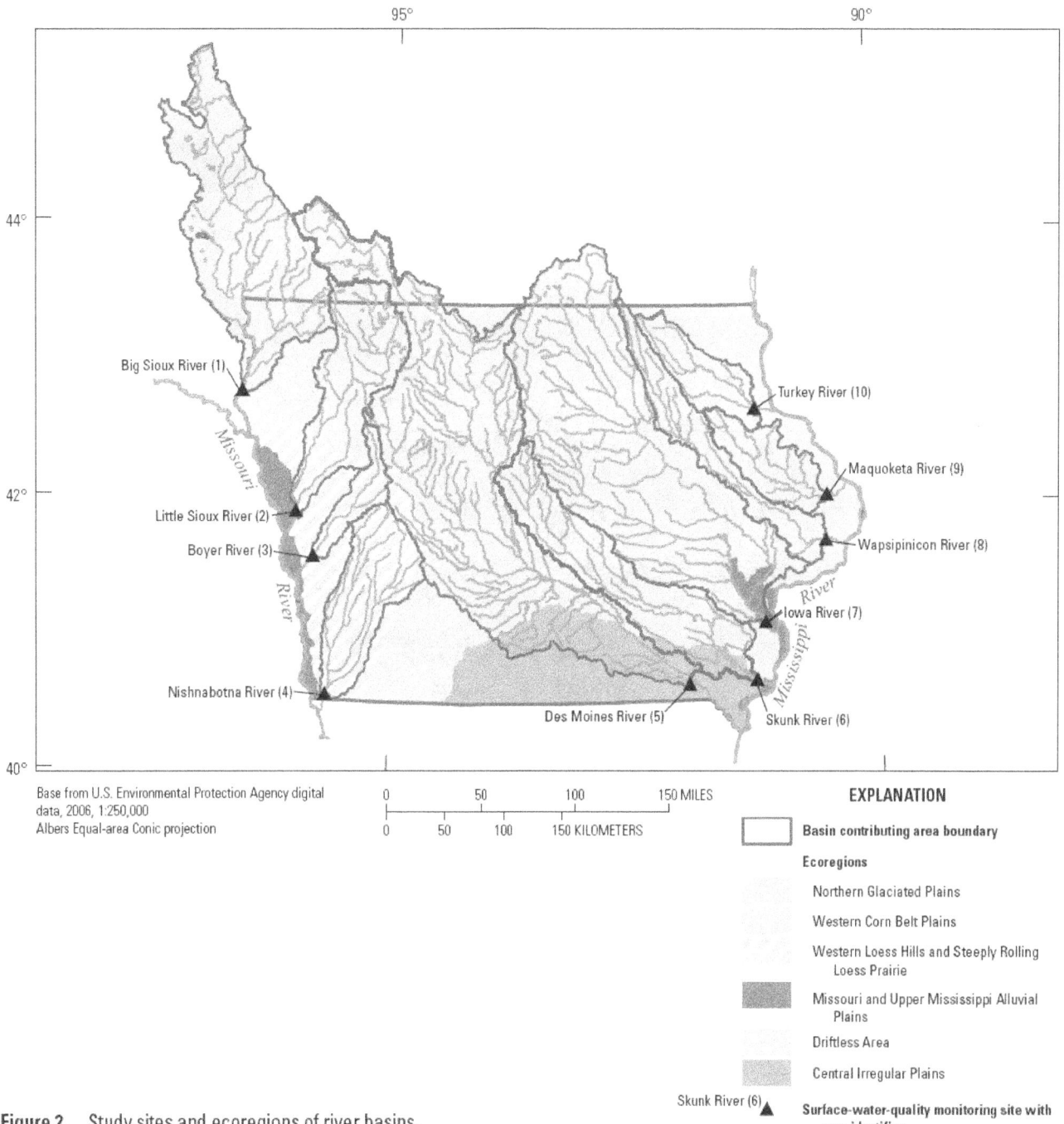

Figure 2. Study sites and ecoregions of river basins.

analytical constituents, and field or analytical methods are outlined in table 2. Analytical results are available in the USGS National Water Information System database (U.S. Geological Survey, 2009). This report contains CAS Registry Numbers®, which is a registered trademark of the American Chemical Society. CAS recommends the verification of the CASRNs through CAS Client Services[SM].

Table 2.   Constituents analyzed at selected major Iowa rivers, water years 2004-2008.

[Water year from October 1 to September 30; --, not available; mm Hg, millimeters of mercury; USGS, U.S. Geological Survey; µS/cm, microsiemens per centimeter;   C, degrees celsius; NTRU, nephelometric turbidity ratio units; mg/L, miligrams per liter; CaCO$_3$, calcium carbonate; SiO$_2$, silicon dioxide; N, nitrogen; P, phosphorus; µg/L, micrograms per liter; DOC, dissolved organic carbon]

Analyte	Parameter code	CAS[1] number	Units	Reference for field or analytical method
Physical properties and sediment				
Barometric pressure	00025	--	mm Hg	USGS, variously dated.
pH	00400	--	standard units	USGS, variously dated.
Specific conductance	00095	--	µS/cm at 25°C	USGS, variously dated.
Water temperature	00010	--	°C	USGS, variously dated.
Turbidity	63676	--	NTRU	USGS, variously dated.
Suspended sediment	80154	--	mg/L	Guy, 1969.
Major ions				
Dissolved oxygen	00300	7782-44-7	mg/L	USGS, variously dated.
Alkalinity, filtered	39086	--	mg/L as CaCO$_3$	USGS, variously dated.
Bicarbonate, filtered	00453	71-52-3	mg/L	USGS, variously dated.
Carbon, total, particulate	00694	7440-44-0	mg/L	Zimmermann, 1997.
Carbonate, filtered	00452	3812-32-6	mg/L	USGS, variously dated.
Chloride, filtered	00940	16887-00-6	mg/L	Fishman and Friedman, 1989.
Carbon, inorganic, particulate	00688	--	mg/L	Zimmermann, 1997.
Silica, filtered	00955	7631-86-9	mg/L as SiO$_2$	Fishman and Friedman, 1989.
Sulfate, filtered	00945	14808-79-8	mg/L	Fishman and Friedman, 1989.
Nutrients and organic carbon				
Ammonia (NH$_3$+NH$_4$), filtered	00608	7664-41-7	mg/L as N	Fishman, 1993.
Nitrate plus nitrite, filtered	00631	--	mg/L as N	Fishman, 1993.
Nitrite, filtered	00613	14797-65-0	mg/L as N	Fishman, 1993.
Orthophosphate, filtered	00671	14265-44-2	mg/L as P	Fishman, 1993.
Total nitrogen, particulate	49570	17778-88-0	mg/L as N	Zimmermann, 1997.
Total phosphorus, unfilterd	00665	7723-14-0	mg/L as P	O'Dell, 1993.
Total nitrogen (NH$_3$+NO$_2$+NO$_3$+organic), filtered	62854	17778-88-0	mg/L as N	Patton and Kryskalla, 2003.
Total nitrogen (NH$_3$+NO$_2$+NO$_3$+organic), unfiltered	62855	17778-88-0	mg/L as N	Patton and Kryskalla, 2003.
Organic carbon, particulate	00689	--	mg/L	Zimmermann, 1997.
Organic carbon, filtered (DOC)	00681	--	mg/L	Brenton and Arnett, 1993.
Algal pigments				
Chlorophyll-*a*	70953	479-61-8	µg/L	Arar and Collins, 1997.
Pheophyton-*a*	62360	603-17-8	µg/L	Arar and Collins, 1997.

[1]This report contains CAS Registry Numbers®, which is a registered trademark of the American Chemical Society. CAS recommends the verification of the CASRNs through CAS Client Services[SM]

Statistical Data Summary

Statistical and graphical summaries of water-quality data were computed using the TIBCO Spotfire S+® statistical package (TIBCO Software Inc., 2008). Because of laboratory reporting conventions, censored values, or values reported as less than the laboratory reporting level (LRL), were re-coded to less than the long-term method detection limit (LT-MDL, typically one-half the reporting level) for the purposes of statistical summary and plotting. This adjustment is needed to account for laboratory data-reporting practices of estimating positive detections above the long-term method detection limit but below the reporting level (Childress and others, 1999). Concentrations are reported as <LRL for samples in which the analyte was either not detected or did not pass identification. Analytes that are detected at concentrations between the LT-MDL and the LRL and that pass identification criteria are reported as estimated. This information-rich laboratory reporting convention retains information about positive low-level detections, which would otherwise be treated simply as a nondetect. Estimated concentrations are reported with a remark code of "E" to indicate greater uncertainty than data reported without the "E" remark code, including values between the LT-MDL and LRL and values with uncertainty affected by matrix effects.

Summary statistics of water-quality data were computed using regression on order statistics (ROS) and adjusted maximum likelihood estimation (AMLE) parametric methods. For handling datasets with values reported below detection limits and for multiple levels of detection, these methods provide better estimation of mean (ROS) and percentiles (AMLE) than either simple substitution (such as one-half detection limit or zero) or ignoring values below detection (Helsel and Hirsch, 2002). Percentiles for pesticide water-quality data were computed using a modification of the Kaplan-Meier nonparametric method (Helsel, 2005). This rank-based nonparametric method is more appropriate to the less frequent detection of many pesticides.

Locally weighted scatterplot smooth (LOWESS) lines are used in figure 3 to conveys similarities and differences among sites, in regards to the overall distribution of the data, and in relation to streamflow. The LOWESS line represents the pattern through the middle of the data, with a weighting function applied for distance from each point along the x-axis and magnitude of residuals in the y-direction (Helsel and Hirsch, 2002).

Load and Yield Estimation

Stream load is defined as the mass of a chemical constituent transported by a stream past one location during a specific period of time, and is expressed in units such as tons per year. Stream yield denotes stream load divided by watershed area expressed in units of mass per year per area, and can be used to compare the relative contributions of constituents from watersheds of different sizes.

Annual loads were estimated using S-LOADEST based on the Fortran LOADEST program described by Runkle and others (2004) and operated in Spotfire S+®. S-LOADEST generates a regression equation between load and variables for streamflow and additional terms specified by the user. Models for each site and constituent were fit using sample data from 2004–2008. The models were applied using daily values for streamflow and additional terms to compute daily loads, which were summed to calculate annual load estimates. Explanatory terms are incorporated as linear additions to the model. S-LOADEST uses AMLE (Cohn, 2005) to correct biases because of censored data and retransformation of load estimated as a log-transformed variable. In cases where the assumption of normality of model residuals could not be met, a least absolute deviations (LAD) method was used.

Standard explanatory terms in S-LOADEST include linear and quadratic terms for streamflow and time plus seasonal terms for time. Streamflow uses daily mean values to correspond to the minimum 1-day time step. Streamflow and time terms are centered so linear and quadratic terms are orthogonal, to eliminate problems associated with collinear explanatory variables (Runkle and others, 2004). Sine and cosine time terms use an annual phase to describe seasonal patterns. A break-point term (BpQ) divides streamflow into two linear segments, and is used in cases where the relation between constituent load and streamflow is better empirically described by separate models at low and high flows.

To improve estimates of storm and snowmelt event transport, additional terms describing streamflow variability and anomalies were evaluated. High-flow events can account for a large part of the total annual load relative to the duration of the events, and rating curve methods tend to underestimate loads at high streamflows (Horowitz, 2003) because the streamflow and time terms in the predefined regression models do not account for event-level factors such as hysteresis or recent events. Hysteresis occurs when concentrations (and thereby loads) are different on rising and falling limbs of an event hydrograph for the same magnitude of streamflow.

Streamflow variability terms were defined as the difference between mean streamflow (Q) on day i and the mean streamflow of the previous k days, given as:

$$dQ_k = \ln Q_i - \sum_{j}^{i-1} \ln Q_j / k \qquad (1)$$

This variability term (dQ) with a 1-day time step (dQ_1) helps describe effects of hysteresis (Wang and Linker, 2008). A term with a 30-day step (dQ_{30}) helps describe effects of sequential events or prolonged event peaks. In some instances, the absolute value of dQ_1 ($|dQ_1|$) better describes loads in the regression model than dQ_1, representing cases where the degree of flashiness of the event was the critical element, rather than hysteresis.

An alternate approach to incorporating streamflow history into load estimation uses time-series terms for streamflow anomalies to describe deviations from average conditions (Vecchia, 2003). The form of the load regression equation with anomalies is given:

$$\ln L = b_o + b_1 A5yr + b_2 A1yr + b_3 A3mo + b_4 HFV \qquad (2)$$

where A5yr, A1yr, A3mo, and HFV sequentially account for variability from long-term average streamflow over different time scales (5-year, 1-year, 3-month, and high-frequency variability, respectively), with regression coefficients, b. Anomalies were computed in Spotfire S+® using daily streamflow data beginning 10 years before the load estimation period. A5yr is the average over a 5-year interval of daily deviations from the long-term average streamflow. A1yr is the average over a 1-year interval of daily deviations between deviations from the long-term average streamflow and A5yr. A3mo and HFV sequentially account for additional variability at finer time scales. Thus, high values for A1yr indicate a wetter year than the previous 5 years; similarly, low values for A3mo indicate a drier season than the previous year. Vecchia (2003) further describes streamflow anomalies.

The process of building and selecting candidate load models using streamflow, time, seasonality, hysteresis, and anomaly terms included automated selection procedures and evaluation of model fit, assumptions on residuals, and correlation of the explanatory variables. First, several procedures were used in tandem to generate a list of candidate models. S-LOADEST models with standard terms were ranked based on Akaike Information Content (AIC) and Schwarz Posterior Probability Criterion (SPPC). Hysteresis terms and streamflow anomalies were evaluated separately with stepwise selection to identify terms that contributed to the standard models. Second, diagnostic tests and plots were considered in selection of the candidate models. Preferred models had low residual variance, residual plots indicating normality and homoskedasticity, and low correlation among explanatory variables indicated by low pairwise correlation and a low multicollinearity statistic (variance inflation factor, VIF). Finally, fit of the best candidate model was verified by comparing observed and predicted daily loads (average measured load from sampled days divided by average estimated load from sampled days) (Stenback and others, 2011). Models for which this ratio was less than one-half or greater than two were not used, and an alternate candidate model was evaluated. Therefore, in this study, models that did not meet these selection criteria were not considered appropriate for estimating loads on unsampled days. For example, of two potential models with different variable selections, the model with a greater residual variance still may have been preferred if diagnostics from the other model indicated a problem with non-normal residuals or an unacceptable ratio of measured/estimated loads.

Because the goal of the modeling was to compute annual loads, models calibrated using certain outlier data points also were not considered appropriate for estimating loads. Outliers were removed from the calibration dataset only if individual data points exhibited undue influence on model parameter estimates and models with an alternate selection of variables could not be used that included all data points. The undue influence of outliers can "pull" the model in one area of the data, resulting in poor model fit in areas critical for the annual load estimates. Exclusion of data does not indicate "bad" data, but rather can indicate an environmental response to something not accounted for in the model (for example, upstream chemical spill). Overall, 1 percent of data were excluded as outliers, with no more than three outliers for any one site and constituent model. Outliers were generally of two categories—high-streamflow event samples and low-streamflow samples with evidence of algae blooms (noted green water color, supersaturated oxygen concentration, for example). Because high streamflow events contribute far more to annual transport than low streamflow, influential outliers at high streamflows were removed only as final course, and the highest sampled flow was never removed from a model calibration set. Low-streamflow influential outliers resulted in models with greater residual errors at high streamflows. Again, high streamflow periods were considered more critical to the objective of estimating annual loads. The potential risk of overestimating concentrations at low streamflow was acknowledged and accepted, as the low-streamflow load contributions to the annual totals were less critical than the high-streamflow load estimates.

Load models calibrated for each site and constituent with the 5-year sample dataset were used to estimate loads for all of 2004 through 2008 water years (WY, defined as October 1 of the previous calendar year through September 30 of the specified year), including standard errors of prediction (SEP) and upper and lower 95-percent confidence limits (U95, L95). Routine sampling began in March 2004. Load estimates for the first one-half of WY2004 required extrapolation below the range of sampled streamflows for all sites except the Maquoketa River near Spragueville (map identifier [ID] 9); the amount of extrapolation needed ranged from a few days when streamflows dipped below the range of sampled flows to the first 5 months at the Big Sioux River at Akron (map ID 1) and Little Sioux River near Turin (map ID 2).

Chemical Concentration, Loads, and Yields in Major Iowa Rivers

Summaries of concentration data, estimated loads, and basin yields are presented in the following section. Concentrations, loads, and yields were analyzed in various ways to describe the data within the context of the landscape and environmental gradients across the State and through time.

Concentrations

Data from routine and high-flow event water samples indicate the effect of streamflow, seasonality, spatial gradients, and basin size. Summary statistics are presented in table 3 for physical properties and concentrations of ions, carbon, nutrients, and suspended sediment. Pesticide summary statistics are presented in table 4 for compounds detected in at least one sample. Pesticides analyzed but not detected through the 5-year study period are listed in table 5.

Streamflow effects on concentrations varied by constituent, but patterns were generally consistent among sites. Streamflow was inversely related to some concentrations, indicating a general dilution effect for pH, alkalinity, specific conductance, chloride, and sulfate (fig. 3A–B). Other constituents such as particulate organic carbon (POC), dissolved organic carbon (DOC), total phosphorus, and suspended sediment had increased concentrations with streamflow (fig. 3C–F). Nitrogen, in the forms of total nitrogen and nitrate plus nitrite, had more complex relations to streamflow, with increasing concentrations through low to average streamflows but decreasing concentrations with streamflows above the 90th percentile for each site. Concentration showed little or no relation to streamflow for silica, particulate inorganic carbon (PIC), or chlorophyll-a. Orthophosphate concentrations also were generally unaffected by or slightly increasing with streamflow, except in the Boyer River (map ID 3), where orthophosphate indicated a strong inverse relation to streamflow (fig. 3E–F).

Because streamflow patterns are seasonal, generally peaking in May and June, patterns in concentration related to streamflow also are seasonal. Seasonal patterns for pH and alkalinity were related to streamflow, with low values evidence of dilution from events common in spring, and high values typical of low stable flows in late summer through winter (fig. 4A). A few constituents, however, showed seasonal patterns not entirely related to streamflow. Algal pigments (chlorophyll-a and pheophytin-a) were affected by late summer algal blooms (chlorophyll-a; fig. 4B). Occasional high concentrations of algal pigments indicated algae blooms in late winter to early spring, particularly in rivers such as the Little Sioux and Des Moines (map ID 5) with upstream lakes or reservoirs. Most detected pesticides tended to have peak concentrations in early spring, shortly after agricultural application (fig. 4C), but concentrations of the nonselective herbicide prometon were greatest from summer through fall (fig. 4D).

Spatial variability of concentrations reflected landform and climate gradients across the State for some constituents. Alkalinity was greater at the northwestern and northeastern sites. Chloride and dissolved organic carbon indicated spatial trends along the border rivers, with increasing concentrations upstream along the Missouri River tributaries (map IDs 4 to 1) and downstream along the Mississippi River tributaries (map IDs 10 to 5). Turbidity, suspended-sediment, and total phosphorus concentrations were greatest in southwestern Iowa streams (map ID 3–4) in basins draining the highly erodible Western Loess Hills. Specific conductance and sulfate

concentrations in the Big Sioux River (map ID 1) were the greatest and most variable of all the sites. Several constituents did not show a pronounced spatial trend, including pH, silica, particulate organic carbon, particulate inorganic carbon, nitrogen (all forms), and algal pigments. The relation between streamflow and concentration also varied by site for POC, orthophosphate, suspended sediment, and turbidity.

Major Ions

Major Iowa rivers are generally alkaline and well-buffered, commonly with calcareous soils and underlying and exposed limestone and dolomite bedrock. For samples from all 10 rivers collected from March 2004 through September 2008, pH was rarely measured below 7.0, and was occasionally measured above 9.0. Mean and median pH were both about 8.1 (table 3). The lowest pH levels occurred during storm and snowmelt events when streamwaters were diluted by rainfall or snowmelt (most common in the spring), and greatest pH levels occurred during long periods of stable streamflow without major precipitation (most common in the summer and fall) (figs. 3A–B, 4A). Sample pH summary statistics were similar for all 10 sites, indicating no major spatial gradients, though discrete sampling data did not allow for analysis of site-specific diurnal patterns. Sample alkalinities commonly ranged from 159 to 242 milligrams per liter as calcium carbonate (mg/L as $CaCO_3$), summarized as the 25th to 75th percentiles for all sites, with an overall range of alkalinities from 42.5 to 356 mg/L as $CaCO_3$ (table 3). Similar to pH, alkalinity measurements were lower from rainfall dilution and greater during long periods of stable streamflow, with greatest alkalinity values in the winter (figs. 3A–B, 4A). Alkalinity exhibited spatial trends among sites, with greater alkalinities in northern sites (map IDs 1, 2, 3, 9, and 10) and rivers draining central basins with alkalinities trending lower from west to east basins (map IDs 4 to 8) (fig. 3A–B).

Specific conductance, chloride, sulfate, and silica concentrations reflect the extent of ion-leaching of natural minerals from the landscape and anthropogenic sources such as point discharges, winter de-icing salts, and agricultural application. Silica concentrations also are affected by biological uptake of diatoms (Wetzel, 1983). For all sites, specific conductance varied from 168 to 1,220 microsiemens per centimeter at 25 degrees Celsius (µS/cm at 25°C) with a median of 574 µS/cm at 25°C and mean of 586 µS/cm at 25°C (table 3). High concentrations generally were observed during low or stable streamflows and low concentrations were the result of rainfall or snowmelt dilution during high streamflows. The Big Sioux River at Akron (site/map ID 1) in the northwest had the greatest and most variable specific conductance values compared with other sites. Chloride concentrations ranged from 2.25 to 86.2 mg/L with a median of 21.6 mg/L and a mean of 23.4 mg/L (table 3). The inverse relation between chloride concentration and streamflow evident at every site was pronounced at some sites, with a coefficient of determination (R^2) from linear regression (Helsel and Hirsch, 2002) between

Table 3.   Statistical summary of select constituents at selected major Iowa rivers, water years 2004–2008.

[Water year from October 1 to September 30; ID, identifier; IQR, interquartile range; parameter code given in brackets, see also table 2; mg/L, milligrams per liter; CaCO$_3$, calcium carbonate; µS/cm, microsiemens per centimeter; <, less than; NTRU, nephelometric turbidity ratio units]

Map ID (fig. 2)	Station name	Number of samples	Number censored	Minimum uncensored	Percentile			IQR	Maximum	Mean
					25	50 (median)	75			
				pH, standard units [00400]						
1	Big Sioux River at Akron, Iowa	58	0	7.2	7.9	8.1	8.3	0.5	8.6	8.1
2	Little Sioux River at Turin, Iowa	58	0	6.9	7.8	8.1	8.3	0.5	8.7	8.0
3	Boyer River at Logan, Iowa	56	0	6.6	7.9	8.2	8.3	0.4	8.7	8.1
4	Nishnabotna River above Hamburg, Iowa	57	0	6.9	7.7	8.0	8.2	0.5	9.0	7.9
5	Des Moines River at Keosauqua, Iowa	55	0	7.0	8.0	8.4	8.6	0.6	9.1	8.3
6	Skunk River at Augusta, Iowa	56	0	7.2	7.8	8.2	8.5	0.7	9.1	8.1
7	Iowa River at Wapello, Iowa	61	0	6.8	7.9	8.2	8.5	0.6	9.1	8.2
8	Wapsipinicon River near De Witt, Iowa	57	0	7.3	7.8	8.1	8.4	0.6	9.0	8.1
9	Maquoketa River near Spragueville, Iowa	56	0	7.3	7.9	8.1	8.3	0.4	8.7	8.1
10	Turkey River at Garber, Iowa	54	0	7.6	7.9	8.1	8.2	0.3	8.5	8.1
	All samples	**568**	**0**	**6.6**	**7.9**	**8.1**	**8.4**	**0.5**	**9.1**	**8.1**
				Alkalinity, mg/L as CaCO$_3$ [39086]						
1	Big Sioux River at Akron, Iowa	57	0	108	192	242	289	97	340	238
2	Little Sioux River at Turin, Iowa	57	0	76.4	187	245	277	90	356	231
3	Boyer River at Logan, Iowa	56	0	83.1	228	255	285	57	340	242
4	Nishnabotna River above Hamburg, Iowa	57	0	73.4	172	212	230	58	326	197
5	Des Moines River at Keosauqua, Iowa	55	0	102	153	179	212	59	286	184
6	Skunk River at Augusta, Iowa	56	0	85.9	145	180	218	73	293	181
7	Iowa River at Wapello, Iowa	59	0	100	130	173	200	70	268	169
8	Wapsipinicon River near De Witt, Iowa	57	0	84.9	119	140	182	63	205	146
9	Maquoketa River near Spragueville, Iowa	54	0	61.7	204	230	251	47	278	218
10	Turkey River at Garber, Iowa	52	0	42.5	204	228	240	36	279	214
	All samples	**560**	**0**	**42.5**	**159**	**206**	**242**	**83**	**356**	**202**

Table 3. Statistical summary of select constituents at selected major Iowa rivers, water years 2004–2008.—Continued

[Water year from October 1 to September 30; ID, identifier; IQR, interquartile range; parameter code given in brackets, see also table 2; mg/L, milligrams per liter; CaCO₃, calcium carbonate; µS/cm, microsiemens per centimeter; <, less than; NTRU, nephlometric turbidity ratio units]

Map ID (fig. 2)	Station name	Number of samples	Number censored	Minimum uncensored	25	50 (median)	75	IQR	Maximum	Mean
						Percentile				
	Specific conductance, µS/cm [00095]									
1	Big Sioux River at Akron, Iowa	58	0	317	804	844	1,004	200	1,220	881
2	Little Sioux River at Turin, Iowa	58	0	168	590	692	731	141	878	650
3	Boyer River at Logan, Iowa	56	0	198	631	684	726	95	907	640
4	Nishnabotna River above Hamburg, Iowa	57	0	237	490	535	560	70	634	502
5	Des Moines River at Keosauqua, Iowa	55	0	348	512	580	656	144	877	585
6	Skunk River at Augusta, Iowa	56	0	252	442	536	620	178	824	530
7	Iowa River at Wapello, Iowa	62	0	312	447	540	584	137	710	524
8	Wapsipinicon River near De Witt, Iowa	57	0	290	373	442	490	117	606	436
9	Maquoketa River near Spragueville, Iowa	56	0	189	544	580	607	63	678	554
10	Turkey River at Garber, Iowa	53	0	228	534	577	590	56	679	552
	All samples	**568**	**0**	**168**	**490**	**574**	**666**	**176**	**1,220**	**586**
	Chloride, mg/L [00940]									
1	Big Sioux River at Akron, Iowa	58	0	10.8	27.8	32.7	38.9	11.1	56.8	33.5
2	Little Sioux River at Turin, Iowa	58	0	5.17	19.6	23.2	25.9	6.3	30.3	22.3
3	Boyer River at Logan, Iowa	57	0	2.25	18.3	22.2	28.2	9.9	86.2	23.7
4	Nishnabotna River above Hamburg, Iowa	58	0	4.28	12.8	15.1	17.8	5.0	25.2	14.9
5	Des Moines River at Keosauqua, Iowa	56	0	10.9	21.5	27.7	36.1	14.6	49.3	28.7
6	Skunk River at Augusta, Iowa	56	0	7.10	19.2	24.8	29.1	9.9	47.3	24.4
7	Iowa River at Wapello, Iowa	60	0	7.84	22.0	29.3	35.7	13.7	45.9	28.4
8	Wapsipinicon River near De Witt, Iowa	57	0	7.32	18.4	20.8	22.6	4.2	29.0	20.6
9	Maquoketa River near Spragueville, Iowa	56	0	4.87	16.4	18.1	20.0	3.6	25.1	18.1
10	Turkey River at Garber, Iowa	55	0	5.77	16.8	19.0	20.2	3.4	26.0	18.5
	All samples	**571**	**0**	**2.25**	**17.4**	**21.6**	**28.2**	**10.8**	**86.2**	**23.4**

Table 3. Statistical summary of select constituents at selected major Iowa rivers, water years 2004–2008.—Continued

[Water year from October 1 to September 30; ID, identifier; IQR, interquartile range; parameter code given in brackets, see also table 2; mg/L, milligrams per liter; CaCO₃, calcium carbonate; µS/cm, microsiemens per centimeter; <, less than; NTRU, nephlometric turbidity ratio units]

Map ID (fig. 2)	Station name	Number of samples	Number censored	Minimum uncensored	Percentile			IQR	Maximum	Mean
					25	50 (median)	75			
			Sulfate, mg/L [00945]							
1	Big Sioux River at Akron, Iowa	58	0	34.6	159	189	204	45	245	177
2	Little Sioux River at Turin, Iowa	58	0	13.6	58.2	70.7	80.5	22.3	98.8	66.9
3	Boyer River at Logan, Iowa	57	0	7.26	37.5	42.7	50.7	13.2	71.6	42.1
4	Nishnabotna River above Hamburg, Iowa	58	0	8.50	24.8	29.0	32.1	7.3	41.8	27.9
5	Des Moines River at Keosauqua, Iowa	56	0	20.0	37.5	53.2	71.2	33.7	109	55.1
6	Skunk River at Augusta, Iowa	56	0	9.38	28.1	33.1	42.6	14.5	76.9	35.6
7	Iowa River at Wapello, Iowa	60	0	8.86	24.3	31.7	36.5	12.2	45.1	30.3
8	Wapsipinicon River near De Witt, Iowa	57	0	7.72	20.7	24.5	27.3	6.6	32.4	23.5
9	Maquoketa River near Spragueville, Iowa	56	0	5.67	23.7	25.3	26.6	2.9	30.5	24.1
10	Turkey River at Garber, Iowa	55	0	7.27	22.6	25.0	27.2	4.6	30.6	24.0
	All samples	**571**	**0**	**5.67**	**25.2**	**32.9**	**54.5**	**29.3**	**245**	**50.9**
			Silica, mg/L [00955]							
1	Big Sioux River at Akron, Iowa	58	0	1.5	10.5	13.5	16.0	5.5	21.7	13.0
2	Little Sioux River at Turin, Iowa	58	0	2.9	11.1	15.7	18.1	7.0	24.9	14.4
3	Boyer River at Logan, Iowa	56	0	4.5	11.3	13.9	16.6	5.3	19.9	13.6
4	Nishnabotna River above Hamburg, Iowa	57	0	6.8	11.6	16.5	15.8	4.2	18.7	13.6
5	Des Moines River at Keosauqua, Iowa	54	0	3.2	10.4	13.4	15.4	5.0	25.8	13.3
6	Skunk River at Augusta, Iowa	54	3	0.2	7.2	12.2	15.1	7.9	21.2	11.4
7	Iowa River at Wapello, Iowa	29	1	1.5	8.0	11.1	13.3	5.3	20.4	10.8
8	Wapsipinicon River near De Witt, Iowa	55	1	0.1	2.6	7.62	10.3	7.7	13.6	6.9
9	Maquoketa River near Spragueville, Iowa	56	0	4.0	4.8	9.96	11.7	6.9	13.6	9.5
10	Turkey River at Garber, Iowa	55	0	2.1	5.4	9.23	10.9	5.5	14.1	8.4
	All samples	**532**	**5**	**0.1**	**8.8**	**11.6**	**14.8**	**6.0**	**25.8**	**11.6**

Table 3. Statistical summary of select constituents at selected major Iowa rivers, water years 2004–2008.—Continued

[Water year from October 1 to September 30; ID, identifier; IQR, interquartile range; parameter code given in brackets, see also table 2; mg/L, milligrams per liter; $CaCO_3$, calcium carbonate; μS/cm, microsiemens per centimeter; <, less than; NTRU, nephelometric turbidity ratio units]

Map ID (fig. 2)	Station name	Number of samples	Number censored	Minimum uncensored	Percentile			IQR	Maximum	Mean
					25	50 (median)	75			
	Organic carbon, particulate, mg/L [00689]									
1	Big Sioux River at Akron, Iowa	58	0	0.38	5.31	7.87	11.5	6.2	32.2	8.53
2	Little Sioux River at Turin, Iowa	58	0	0.36	3.72	7.52	12.2	8.5	108	13.6
3	Boyer River at Logan, Iowa	56	0	0.30	1.42	3.41	9.41	7.99	285	17.6
4	Nishnabotna River above Hamburg, Iowa	57	0	0.43	1.51	5.24	12.3	10.8	137	13.0
5	Des Moines River at Keosauqua, Iowa	54	0	0.41	1.52	2.38	3.83	2.31	51.4	4.55
6	Skunk River at Augusta, Iowa	56	0	0.44	3.11	4.35	6.88	3.77	35.4	6.16
7	Iowa River at Wapello, Iowa	41	0	0.48	2.66	4.71	8.06	5.4	16.1	5.34
8	Wapsipinicon River near De Witt, Iowa	55	0	0.27	1.76	4.64	9.05	7.29	19.9	5.96
9	Maquoketa River near Spragueville, Iowa	56	0	0.28	1.16	2.48	4.30	3.14	56.4	5.02
10	Turkey River at Garber, Iowa	55	1	0.27	0.72	1.57	3.82	3.11	108	6.33
	All samples	**543**	**1**	**0.27**	**1.57**	**4.00**	**8.53**	**6.96**	**285**	**8.77**
	Inorganic carbon, particulate, mg/L [00688]									
1	Big Sioux River at Akron, Iowa	58	17	0.04	0.05	0.27	1.04	0.99	8.44	0.90
2	Little Sioux River at Turin, Iowa	58	21	0.02	<0.04	0.25	0.72	0.69	4.88	0.63
3	Boyer River at Logan, Iowa	56	30	0.03	<0.04	0.04	0.18	0.17	7.49	0.49
4	Nishnabotna River above Hamburg, Iowa	57	31	0.03	<0.04	0.04	0.19	0.17	8.08	0.31
5	Des Moines River at Keosauqua, Iowa	54	42	0.03	<0.04	<0.04	0.03	0.03	6.37	0.28
6	Skunk River at Augusta, Iowa	53	34	0.02	<0.04	0.02	0.12	0.11	7.00	0.27
7	Iowa River at Wapello, Iowa	41	24	0.02	0.03	0.13	0.67	0.65	5.67	0.68
8	Wapsipinicon River near De Witt, Iowa	55	21	0.02	<0.04	0.15	0.56	0.54	5.05	0.73
9	Maquoketa River near Spragueville, Iowa	56	40	0.03	<0.04	0.02	0.08	0.07	0.96	0.11
10	Turkey River at Garber, Iowa	55	33	0.02	<0.04	0.02	0.24	0.23	26.3	0.98
	All samples	**543**	**283**	**0.02**	**0.01**	**0.05**	**0.29**	**0.28**	**26.3**	**0.54**

Table 3.   Statistical summary of select constituents at selected major Iowa rivers, water years 2004–2008.—Continued

[Water year from October 1 to September 30; ID, identifier; IQR, interquartile range; parameter code given in brackets, see also table 2; mg/L, milligrams per liter; CaCO₃, calcium carbonate; μS/cm, microsiemens per centimeter; <, less than; NTRU, nephlometric turbidity ratio units]

Map ID (fig. 2)	Station name	Number of samples	Number censored	Minimum uncensored	Percentile 25	50 (median)	75	IQR	Maximum	Mean
				Organic carbon, dissolved, mg/L [00681]						
1	Big Sioux River at Akron, Iowa	58	0	2.4	3.8	4.6	6.4	2.6	17.4	5.5
2	Little Sioux River at Turin, Iowa	58	0	2.3	2.9	3.4	4.2	1.4	15.8	3.9
3	Boyer River at Logan, Iowa	56	0	1.8	2.4	2.8	4.3	1.9	15.1	3.8
4	Nishnabotna River above Hamburg, Iowa	57	0	1.6	2.2	2.6	3.9	1.6	12.3	3.3
5	Des Moines River at Keosauqua, Iowa	54	0	3.3	4.1	4.4	5.1	1.0	9.0	4.8
6	Skunk River at Augusta, Iowa	53	0	2.4	3.0	3.9	5.2	2.2	8.9	4.3
7	Iowa River at Wapello, Iowa	42	0	2.5	3.5	4.0	4.9	1.4	22.9	5.4
8	Wapsipinicon River near De Witt, Iowa	55	0	1.7	2.4	2.9	3.9	1.5	8.3	3.3
9	Maquoketa River near Spragueville, Iowa	56	0	9.4	1.7	2.3	3.0	1.3	11.7	2.7
10	Turkey River at Garber, Iowa	55	0	0.9	1.7	2.3	3.1	1.5	11.8	2.8
	All samples	**544**	**0**	**0.9**	**2.5**	**3.5**	**4.5**	**2.1**	**22.9**	**3.9**
				Total nitrogen, mg/L [49570 plus 62854, or 62855]						
1	Big Sioux River at Akron, Iowa	56	0	0.39	6.03	7.31	9.07	3.04	11.2	7.22
2	Little Sioux River at Turin, Iowa	58	0	2.06	6.22	8.95	10.3	4.1	16.1	8.68
3	Boyer River at Logan, Iowa	57	0	3.42	7.48	9.34	11.5	4.0	37.8	10.1
4	Nishnabotna River above Hamburg, Iowa	58	0	2.22	4.93	7.38	9.25	4.32	22.0	7.47
5	Des Moines River at Keosauqua, Iowa	56	0	2.40	4.72	6.42	9.07	4.35	12.9	6.74
6	Skunk River at Augusta, Iowa	56	0	0.71	4.41	7.12	9.87	5.46	15.1	7.00
7	Iowa River at Wapello, Iowa	60	0	2.25	5.58	6.64	9.16	3.58	14.0	7.44
8	Wapsipinicon River near De Witt, Iowa	55	0	1.40	5.01	7.21	8.76	3.75	15.4	7.37
9	Maquoketa River near Spragueville, Iowa	56	0	3.79	6.29	7.91	9.51	3.22	16.6	7.99
10	Turkey River at Garber, Iowa	55	0	4.01	5.99	7.39	10.0	4.0	15.0	7.82
	All samples	**569**	**0**	**0.39**	**5.67**	**7.49**	**9.79**	**4.12**	**37.8**	**7.79**

Table 3. Statistical summary of select constituents at selected major Iowa rivers, water years 2004–2008.—Continued

[Water year from October 1 to September 30; ID, identifier; IQR, interquartile range; parameter code given in brackets, see also table 2; mg/L, milligrams per liter; CaCO₃, calcium carbonate; μS/cm, microsiemens per centimeter; <, less than; NTRU, nephlometric turbidity ratio units]

Map ID (fig. 2)	Station name	Number of samples	Number censored	Minimum uncensored	25	Percentile 50 (median)	75	IQR	Maximum	Mean
				Nitrate plus nitrite, mg/L [00631]						
1	Big Sioux River at Akron, Iowa	58	0	0.86	4.45	5.84	7.69	3.24	9.26	5.66
2	Little Sioux River at Turin, Iowa	58	0	0.30	4.74	7.42	8.70	3.96	12.3	6.70
3	Boyer River at Logan, Iowa	56	0	1.50	5.45	7.56	9.65	4.20	13.2	7.60
4	Nishnabotna River above Hamburg, Iowa	58	0	1.66	3.39	5.40	7.46	4.07	11.5	5.59
5	Des Moines River at Keosauqua, Iowa	56	0	1.02	3.80	4.85	7.40	3.60	11.9	5.55
6	Skunk River at Augusta, Iowa	56	6	0.15	3.32	5.81	8.41	5.09	14.6	5.71
7	Iowa River at Wapello, Iowa	60	0	0.23	4.42	5.49	7.88	3.46	12.9	6.00
8	Wapsipinicon River near De Witt, Iowa	57	1	0.12	4.19	6.18	7.57	3.38	13.6	6.03
9	Maquoketa River near Spragueville, Iowa	56	0	2.40	5.06	6.80	8.20	3.14	15.0	6.84
10	Turkey River at Garber, Iowa	55	0	2.33	5.07	6.78	7.90	2.83	12.8	6.77
	All samples	**571**	**7**	**0.12**	**4.32**	**6.15**	**8.23**	**3.91**	**15.0**	**6.24**
				Ammonia, mg/L as N [00608]						
1	Big Sioux River at Akron, Iowa	58	22	0.006	<0.01	0.021	0.094	0.088	1.14	0.113
2	Little Sioux River at Turin, Iowa	58	23	0.006	<0.01	0.019	0.058	0.052	1.04	0.088
3	Boyer River at Logan, Iowa	57	23	0.009	<0.01	0.022	0.068	0.061	0.883	0.109
4	Nishnabotna River above Hamburg, Iowa	58	25	0.007	<0.01	0.020	0.098	0.092	0.579	0.075
5	Des Moines River at Keosauqua, Iowa	56	30	0.006	<0.01	0.012	0.044	0.040	0.622	0.053
6	Skunk River at Augusta, Iowa	56	27	0.005	<0.01	0.013	0.046	0.042	0.630	0.070
7	Iowa River at Wapello, Iowa	60	34	0.006	<0.01	<0.01	0.031	0.028	0.684	0.054
8	Wapsipinicon River near De Witt, Iowa	57	28	0.007	<0.01	0.013	0.037	0.032	0.841	0.050
9	Maquoketa River near Spragueville, Iowa	56	28	0.006	<0.01	0.014	0.046	0.042	1.31	0.081
10	Turkey River at Garber, Iowa	55	32	0.009	<0.01	<0.01	0.040	0.038	1.06	0.059
	All samples	**571**	**270**	**0.005**	**<0.01**	**0.015**	**0.052**	**0.048**	**1.31**	**0.075**

Table 3.   Statistical summary of select constituents at selected major Iowa rivers, water years 2004–2008.—Continued

[Water year from October 1 to September 30; ID, identifier; IQR, interquartile range; parameter code given in brackets; see also table 2; mg/L, milligrams per liter; $CaCO_3$, calcium carbonate; µS/cm, microsiemens per centimeter; <, less than; NTRU, nephlometric turbidity ratio units]

Map ID (fig. 2)	Station name	Number of samples	Number censored	Minimum uncensored	Percentile			IQR	Maximum	Mean
					25	50 (median)	75			
	Total phosphorus, mg/L [00665]									
1	Big Sioux River at Akron, Iowa	58	0	0.159	0.305	0.362	0.499	0.194	1.43	0.462
2	Little Sioux River at Turin, Iowa	58	0	0.060	0.184	0.286	0.515	0.331	5.16	0.524
3	Boyer River at Logan, Iowa	57	0	0.368	0.496	0.688	1.00	0.50	7.77	1.27
4	Nishnabotna River above Hamburg, Iowa	58	0	0.118	0.236	0.452	0.875	0.639	9.41	0.944
5	Des Moines River at Keosauqua, Iowa	55	0	0.124	0.208	0.278	0.345	0.137	1.02	0.304
6	Skunk River at Augusta, Iowa	55	0	0.054	0.197	0.327	0.459	0.262	1.15	0.381
7	Iowa River at Wapello, Iowa	60	0	0.035	0.267	0.312	0.372	0.105	0.868	0.324
8	Wapsipinicon River near De Witt, Iowa	57	0	0.055	0.143	0.209	0.265	0.122	0.681	0.236
9	Maquoketa River near Spragueville, Iowa	56	0	0.081	0.145	0.195	0.308	0.163	2.72	0.342
10	Turkey River at Garber, Iowa	55	0	0.023	0.065	0.112	0.208	0.143	1.80	0.260
	All samples	**569**	**0**	**0.023**	**0.191**	**0.303**	**0.486**	**0.295**	**9.41**	**0.507**
	Orthophosphate, mg/L [00671]									
1	Big Sioux River at Akron, Iowa	58	0	0.003	0.037	0.153	0.218	0.181	0.587	0.160
2	Little Sioux River at Turin, Iowa	58	7	0.004	0.050	0.071	0.120	0.071	0.373	0.090
3	Boyer River at Logan, Iowa	57	0	0.054	0.260	0.347	0.570	0.310	1.22	0.431
4	Nishnabotna River above Hamburg, Iowa	58	0	0.057	0.098	0.130	0.158	0.060	0.208	0.131
5	Des Moines River at Keosauqua, Iowa	56	1	0.005	0.086	0.147	0.175	0.089	0.316	0.147
6	Skunk River at Augusta, Iowa	56	2	0.005	0.058	0.112	0.170	0.112	0.365	0.115
7	Iowa River at Wapello, Iowa	60	3	0.004	0.054	0.123	0.166	0.112	0.445	0.115
8	Wapsipinicon River near De Witt, Iowa	57	12	0.003	0.007	0.036	0.074	0.067	0.391	0.051
9	Maquoketa River near Spragueville, Iowa	56	2	0.008	0.049	0.078	0.134	0.086	0.684	0.105
10	Turkey River at Garber, Iowa	55	6	0.006	0.016	0.042	0.090	0.074	0.663	0.065
	All samples	**571**	**33**	**0.003**	**0.050**	**0.104**	**0.172**	**0.123**	**1.22**	**0.140**

Table 3. Statistical summary of select constituents at selected major Iowa rivers, water years 2004–2008.—Continued

[Water year from October 1 to September 30; ID, identifier; IQR, interquartile range; parameter code given in brackets, see also table 2; CaCO₃, calcium carbonate; μS/cm, microsiemens per centimeter; <, less than; NTRU, nephlometric turbidity ratio units]

Map ID (fig. 2)	Station name	Number of samples	Number censored	Minimum uncensored	Percentile			IQR	Maximum	Mean
					25	50 (median)	75			
	Suspended sediment, mg/L [80154]									
1	Big Sioux River at Akron, Iowa	57	0	32	106	181	322	216	1,180	245
2	Little Sioux River at Turin, Iowa	56	0	26	125	247	588	463	4,900	670
3	Boyer River at Logan, Iowa	56	0	5	57.5	260	701	644	22,600	1,600
4	Nishnabotna River above Hamburg, Iowa	57	0	19	126	321	1,400	1,270	8,700	1,110
5	Des Moines River at Keosauqua, Iowa	56	0	3	24.5	73.5	149	125	3,520	231
6	Skunk River at Augusta, Iowa	55	0	9	63	128	380	317	2,520	306
7	Iowa River at Wapello, Iowa	58	0	12	67	130	201	134	596	159
8	Wapsipinicon River near De Witt, Iowa	55	0	17	64	113	221	157	1,760	208
9	Maquoketa River near Spragueville, Iowa	54	0	24	85	126	261	176	2,570	300
10	Turkey River at Garber, Iowa	54	0	29	74	116	191	117	1,880	272
	All samples	**558**	**0**	**3**	**74**	**138**	**348**	**274**	**22,600**	**512**
	Turbidity, NTRU [63676]									
1	Big Sioux River at Akron, Iowa	47	0	2.8	19	52	69	49	570	73
2	Little Sioux River at Turin, Iowa	46	0	2.5	25	60	120	95	1,360	150
3	Boyer River at Logan, Iowa	45	0	2.9	17	55	140	120	8,050	440
4	Nishnabotna River above Hamburg, Iowa	46	0	4.3	19	99	240	220	2,020	280
5	Des Moines River at Keosauqua, Iowa	45	0	2.3	9.9	22	39	29	1,380	68
6	Skunk River at Augusta, Iowa	45	0	2.2	17	60	110	93	660	100
7	Iowa River at Wapello, Iowa	29	0	3.9	18	37	61	43	210	50
8	Wapsipinicon River near De Witt, Iowa	45	0	2.6	14	31	49	35	110	37
9	Maquoketa River near Spragueville, Iowa	45	0	3.6	8.7	24	44	35	1,510	78
10	Turkey River at Garber, Iowa	45	0	2.3	4.1	12	23	18	780	55
	All samples	**438**	**0**	**2.2**	**13**	**33**	**90**	**77**	**8,050**	**140**

Table 3.   Statistical summary of select constituents at selected major Iowa rivers, water years 2004–2008.—Continued

[Water year from October 1 to September 30; ID, identifier; IQR, interquartile range; parameter code given in brackets, see also table 2; mg/L, milligrams per liter; CaCO$_3$, calcium carbonate; µS/cm, microsiemens per centimeter; <, less than; NTRU, nephelometric turbidity ratio units]

Map ID (fig. 2)	Station name	Number of samples	Number censored	Minimum uncensored	Percentile 25	Percentile 50 (median)	Percentile 75	IQR	Maximum	Mean
				Chlorophyll-a, µg/L [70953]						
1	Big Sioux River at Akron, Iowa	57	0	0.9	10.6	42.3	139	128	384	78.8
2	Little Sioux River at Turin, Iowa	58	0	0.6	7.7	19.6	62.3	54.6	328	47.6
3	Boyer River at Logan, Iowa	56	1	1.3	5.6	9.5	16.3	10.8	100	16.2
4	Nishnabotna River above Hamburg, Iowa	57	0	1.0	3.9	7.1	16.4	12.5	91.5	16.2
5	Des Moines River at Keosauqua, Iowa	54	0	1.3	7.0	21.3	36.2	29.2	139	28.7
6	Skunk River at Augusta, Iowa	54	0	1.1	4.9	14.5	51.7	46.8	245	43.6
7	Iowa River at Wapello, Iowa	35	0	1.6	10.8	35.8	99.3	88.5	255	62.6
8	Wapsipinicon River near De Witt, Iowa	55	0	0.9	4.8	26.8	106	101	367	65.1
9	Maquoketa River near Spragueville, Iowa	56	0	0.3	4.0	11.6	32.3	28.3	81.3	20.1
10	Turkey River at Garber, Iowa	55	0	0.7	3.0	6.3	16.8	13.8	94.8	12.4
	All samples	**537**	**1**	**0.3**	**5.4**	**13.3**	**43.0**	**37.6**	**384**	**38.5**
				Pheophytin-a, µg/L [62360]						
1	Big Sioux River at Akron, Iowa	57	0	0.9	7.4	16.8	55.3	47.9	128	30.9
2	Little Sioux River at Turin, Iowa	58	0	0.4	4.0	8.5	24.7	20.8	112	18.0
3	Boyer River at Logan, Iowa	56	1	0.8	2.9	5.4	9.9	7.0	96.4	11.5
4	Nishnabotna River above Hamburg, Iowa	57	0	0.6	3.1	4.6	9.4	6.3	59.7	8.8
5	Des Moines River at Keosauqua, Iowa	54	0	1.4	3.6	8.6	17.6	14.1	76.5	13.8
6	Skunk River at Augusta, Iowa	54	0	0.9	3.7	11.4	31.9	28.2	90.2	21.5
7	Iowa River at Wapello, Iowa	35	0	1.1	6.4	18.1	47.8	41.4	123	33.4
8	Wapsipinicon River near De Witt, Iowa	55	0	0.4	4.3	12.0	46.3	42.0	181	27.4
9	Maquoketa River near Spragueville, Iowa	56	0	0.5	2.5	6.6	17.1	14.6	52.8	10.7
10	Turkey River at Garber, Iowa	55	0	0.8	2.5	5.0	12.6	10.1	87.2	10.8
	All samples	**537**	**1**	**0.4**	**3.5**	**8.3**	**22.5**	**19.1**	**181**	**18.1**

Table 4. Statistical summary of pesticides at selected major Iowa rivers, water years 2004–2008.

[Water year from October 1 to September 30; µg/L, micrograms per liter; lrl, lab reporting limit; --, not applicable]

Analyte	Parameter code	CAS number	Number of samples	Number censored	Percentile (µg/L)				Maximum, (µg/L)	Censoring levels (1/2 lrl, in µg/L)
					25	50 (median)	75	90		
Acetochlor	49260	34256-82-1	564	77	0.006	0.015	0.062	0.418	6.23	0.003–0.022
Alachlor	46342	15972-60-8	564	447	0.002	0.002	0.002	0.007	0.26	0.002–0.010
2,6-Diethylaniline	82660	579-66-8	560	558	--	--	--	--	0.002	0.001–0.003
Atrazine	39632	1912-24-9	560	0	0.061	0.101	0.266	1.38	41	--
2-Chloro-4-isopropylamino-6-amino-s-triazine {CIAT}	04040	6190-65-4	560	0	0.027	0.048	0.077	0.132	0.848	--
Butylate	04028	2008-41-5	536	534	--	--	--	--	0.038	0.001–0.006
Carbaryl	82680	63-25-2	560	551	--	--	--	--	0.032	0.020–0.030
Carbofuran	82674	1563-66-2	556	542	--	--	--	--	0.736	0.009–0.020
Chlorpyrifos	38933	2921-88-2	560	507	0.001	0.002	0.002	0.002	0.14	0.002–0.009
Cyanazine	04041	21725-46-2	556	506	0.01	0.01	0.01	0.01	0.52	0.01–0.03
Dacthal	82682	1861-32-1	560	554	--	--	--	--	0.003	0.002
Diazinon	39572	333-41-5	560	554	--	--	--	0.002	0.024	0.002
Dieldrin	39381	60-57-1	560	554	--	0.002	0.003	0.003	0.053	0.004–0.100
EPTC	82668	759-94-4	556	528	--	--	--	0.001	0.025	0.001–0.040
Desulfinylfipronil amide	62169	--	560	558	--	--	--	--	0.008	0.014
Fipronil sulfide	62167	120067-83-6	560	553	--	--	--	--	0.012	0.006
Fipronil sulfone	62168	120068-36-2	560	545	--	--	--	--	0.010	0.012
Desulfinylfipronil	62170	--	560	536	--	--	--	--	0.008	0.006
Fipronil	62166	120068-37-3	560	508	--	--	--	--	0.031	0.008–0.010
Fonofos	04095	944-22-9	560	559	--	--	--	--	0.024	0.002–0.005
Metolachlor	39415	51218-45-2	564	0	0.026	0.052	0.123	0.451	6.92	--
Metribuzin	82630	21087-64-9	560	494	--	--	0.003	0.006	0.139	0.003–0.014
Napropamide	82684	15299-99-7	536	535	--	--	--	--	0.006	0.004–0.009
p,p'-DDE	34653	72-55-9	536	535	--	--	--	--	0.005	0.002–0.008
Pendimethalin	82683	40487-42-1	560	540	--	0.006	0.006	0.006	0.003	0.006–0.013
Prometon	04037	1610-18-0	560	241	0.003	0.005	0.01	0.02	0.28	0.002–0.008
Propachlor	04024	1918-16-7	553	551	--	--	--	--	0.0317	0.003–0.0125
Propanil	82679	709-98-8	556	555	--	--	--	--	0.026	0.003–0.050
Simazine	04035	122-34-9	560	327	0.002	0.002	0.008	0.018	0.188	0.002–0.005
Tebuthiuron	82670	34014-18-1	560	549	--	--	--	--	0.044	0.008–0.012
Terbacil	82665	5902-51-2	536	533	--	--	--	--	0.057	0.009–0.020
Trifluralin	82661	1582-09-8	560	531	--	--	--	0.003	0.077	0.003–0.004

[1]This report contains CAS Registry Numbers®, which is a registered trademark of the American Chemical Society. CAS recommends the verification of the CASRNs through CAS Client Services℠.

chloride concentration and log-streamflow as high as 0.79. Observed chloride concentrations tended to be greater and more variable at some sites, with a general increase in concentrations upstream among tributaries along the Missouri River (map IDs 4 to 1) and downstream among tributaries along the Mississippi River (map IDs 10 to 5) (fig. 3A–B). Sulfate concentrations varied from 5.67 to 245 mg/L with a median of 32.9 mg/L and a mean of 50.9 mg/L (table 3). An inverse

relation between sulfate and streamflow was observed for each site, with the greatest concentrations and widest range measured at the Big Sioux River at Akron (map ID 1, fig. 3A–B). Silica concentrations ranged from below detection (less than 0.2) to 25.8 mg/L with a median and mean of 11.6 mg/L (table 3). Unlike chloride and sulfate, silica concentrations were not correlated with streamflow and did not exhibit spatial patterns among the 10 sites (fig. 3A–B).

Table 5.   Pesticides analyzed but not detected at selected major Iowa rivers, water years 2004–2008.

[Water year from October 1 to September 30; µg/L, micrograms per liter; lrl, lab reporting level]

Analyte	Parameter code	CAS number	Number of samples	Censoring levels (1/2 lrl, in µg/L)
alpha-HCH	34253	319-84-6	536	0.001–0.002
Azinphos-methyl	82686	86-50-0	560	0.025–0.060
Benfluralin	82673	1861-40-1	560	0.002–0.005
cis-Permethrin	82687	61949-76-6	560	0.003–0.005
Disulfoton	82677	298-04-4	556	0.01–0.02
Ethalfluralin	82663	55283-68-6	536	0.004
Ethoprophos	82672	13194-48-4	556	0.002–0.018
Lindane	39341	58-89-9	536	0.002–0.003
Linuron	82666	330-55-2	536	0.018–0.030
Malathion	39532	121-75-5	560	0.008–0.014
Parathion-methyl	82667	298-00-0	560	0.004–0.008
Molinate	82671	2212-67-1	556	0.001–0.002
Parathion	39542	56-38-2	536	0.005
Pebulate	82669	1114-71-2	536	0.002–0.004
Phorate	82664	298-02-2	560	0.006–0.028
Propyzamide	82676	23950-58-5	560	0.002–0.005
Propargite	82685	2312-35-8	556	0.01–0.02
Terbufos	82675	13071-79-9	560	0.006–0.009
Thiobencarb	82681	28249-77-6	556	0.005
Tri-allate	82678	2303-17-5	536	0.001–0.003

[1]This report contains CAS Registry Numbers®, which is a registered trademark of the American Chemical Society. CAS recommends the verification of the CASRNs through CAS Client Services[SM].

Carbon

Carbon in streamwater was dominated by inorganic ions such as bicarbonate, discussed previously; and particulate carbon concentrations were generally greater than DOC concentrations (fig. 5). POC concentrations ranged from below detection (less than 0.12) to 285 mg/L with a median of 4.00 mg/L and mean of 8.77 mg/L (table 3). At some sites POC concentration and streamflow were not correlated, whereas other sites were strongly correlated but only through a range of flows (fig. 3C–D). For example, POC concentrations at the Little Sioux River near Turin (map ID 2) and the Maquoketa River near Spragueville (map ID 9) indicate no correlation with streamflow less than 1,500 ft³/s, and a strong positive relation for streamflow greater than 1,500 ft³/s, whereas POC concentrations in the Iowa River (map ID 7) indicated no relation to streamflow. The differences in POC concentrations among sites did not reveal any distinct spatial trends. Particulate inorganic carbon (PIC) was frequently observed below detection (less than 0.04 or 0.12 mg/L) with a maximum concentration of 26.3 mg/L, a median of 0.05 mg/L, and a mean of 0.54 mg/L (table 3). No consistent pattern was observed between PIC concentration and streamflow or spatially among the sites (fig. 3C–D). DOC concentrations ranged from 0.9 to 22.9 mg/L with a median of 3.5 mg/L and mean of 3.9 mg/L (table 3). The correlation between DOC concentration and streamflow was positive for most sites and most ranges of streamflow (fig. 3C–D). Spatially, DOC concentrations trended downward from upstream to downstream Missouri River tributaries and upward in upstream to downstream Mississippi River tributaries.

Nitrogen

Nitrogen in Iowa streams affects in-stream aquatic environments, designated human uses, and the water quality of downstream rivers and the Gulf of Mexico. Total nitrogen (TN), was computed as the sum of total dissolved nitrogen (filtered, by alkaline persulfate digestion, parameter code 62854; table 2) plus particulate nitrogen (parameter code 49570); total nitrogen by alkaline persulfate digestion of unfiltered water (parameter code 62855) was used where the separate summation could not be made. All nitrogen species concentrations are reported as milligrams per liter as nitrogen. Nitrate is typically a large component of the total nitrogen in Iowa rivers (median about 85 percent). Nitrite and ammonia concentrations account for very little of the total, however, the in-stream deleterious consequences of these dissolved species occur at different levels.

Nitrate and nitrite numerical regulations do not apply to the Iowa rivers selected for this study, because the stream reaches are not used to supply drinking water, and stream nutrient criteria have not yet been proposed for the protection of aquatic life in Iowa. The maximum contaminant level (MCL) for surface waters designated for Iowa public drinking supplies is 10 mg/L for nitrate and 1 mg/L for nitrite (Iowa

Environmental Protection Commission [567], 2002). The draft nitrate criteria for the protection of aquatic life in warm-water lakes and streams in Minnesota is 4.9 mg/L as a 4-day chronic criteria (Monson, 2010). Concentrations lower than these criteria also can contribute to nuisance algae growth, substantial nitrogen delivery to the Gulf of Mexico, and toxicity for sensitive aquatic life. Nitrate concentrations were equal to or greater than the nitrate MCL in at least one sample at all sites except the Big Sioux River at Akron (map ID 1), and 11 percent of all samples had at least 10 mg/L nitrate. Nitrate concentrations at or above the proposed Minnesota aquatic life criteria occurred in 68 percent of samples at all sites, with 50 to 84 percent of samples at individual sites exceeding 4.9 mg/L. The maximum observed nitrite concentration at the study sites was 0.164 mg/L, well below the MCL.

Criteria for ammonia vary with temperature and pH, and all observed concentrations met the acute and chronic criteria. Though the chronic criterion applies to a 30-day average concentration, the 30-day criterion is unlikely to be exceeded if 95 percent of discrete samples do not exceed the chronic criteria (U.S. Environmental Protection Agency, 1999). Ammonia concentrations also met the 2009 proposed acute ammonia criteria, which accounts for more sensitive mussel tolerances along with early life stages of fish (U.S. Environmental Protection Agency, 2009a). Though the 30-day average could not be evaluated directly for the proposed chronic criteria for these data, one sample in each of the Big Sioux (map ID 1), Boyer (map ID 3), and Maquoketa (map ID 9) Rivers exceeded the more sensitive proposed chronic criteria, indicating the possibility of exceeding the proposed chronic criteria.

Total nitrogen concentrations ranged from 0.39 mg/L to 37.8 mg/L distributed around a median 7.49 mg/L and a mean 7.79 mg/L (table 3). The relation between TN and streamflow varied across ranges of streamflow; increasing from low to average streamflows, then plateauing at many sites before showing effects of dilution at high streamflows (above the 90th percentile, fig. 3C–D, one outlier [Boyer River, 37.8 mg/L] is above the plotted scale for total nitrogen). TN concentration statistics did not vary substantially among the 10 study sites. Nitrate concentrations ranged from below detection (less than 0.06 mg/L) to 15.0 mg/L, with a median 6.15 mg/L and a mean 6.24 mg/L (table 3). Because nitrate is commonly a large portion of TN, both constituents relate similarly to streamflow and among sites (fig. 3C–D). Nitrite was not measured to be a substantial component of nitrate, (measured as nitrate plus nitrite), with nitrite concentrations exceeding the laboratory reporting level for nitrate plus nitrite of 0.04 mg/L in only 18.6 percent of samples. The maximum observed value for nitrite was 0.164 mg/L, well below the nitrate concentrations. Measurable ammonia concentrations ranged from 0.0051 (estimated value above the long-term method detection limit but below the laboratory reporting level, [Childress and others, 2009]) to 1.31 mg/L, which accounts for ammonia (NH_3) and the more prevalent and ecologically benign ammonium ion (NH_4^+). Ammonia was undetected in nearly 50 percent of samples, with reporting

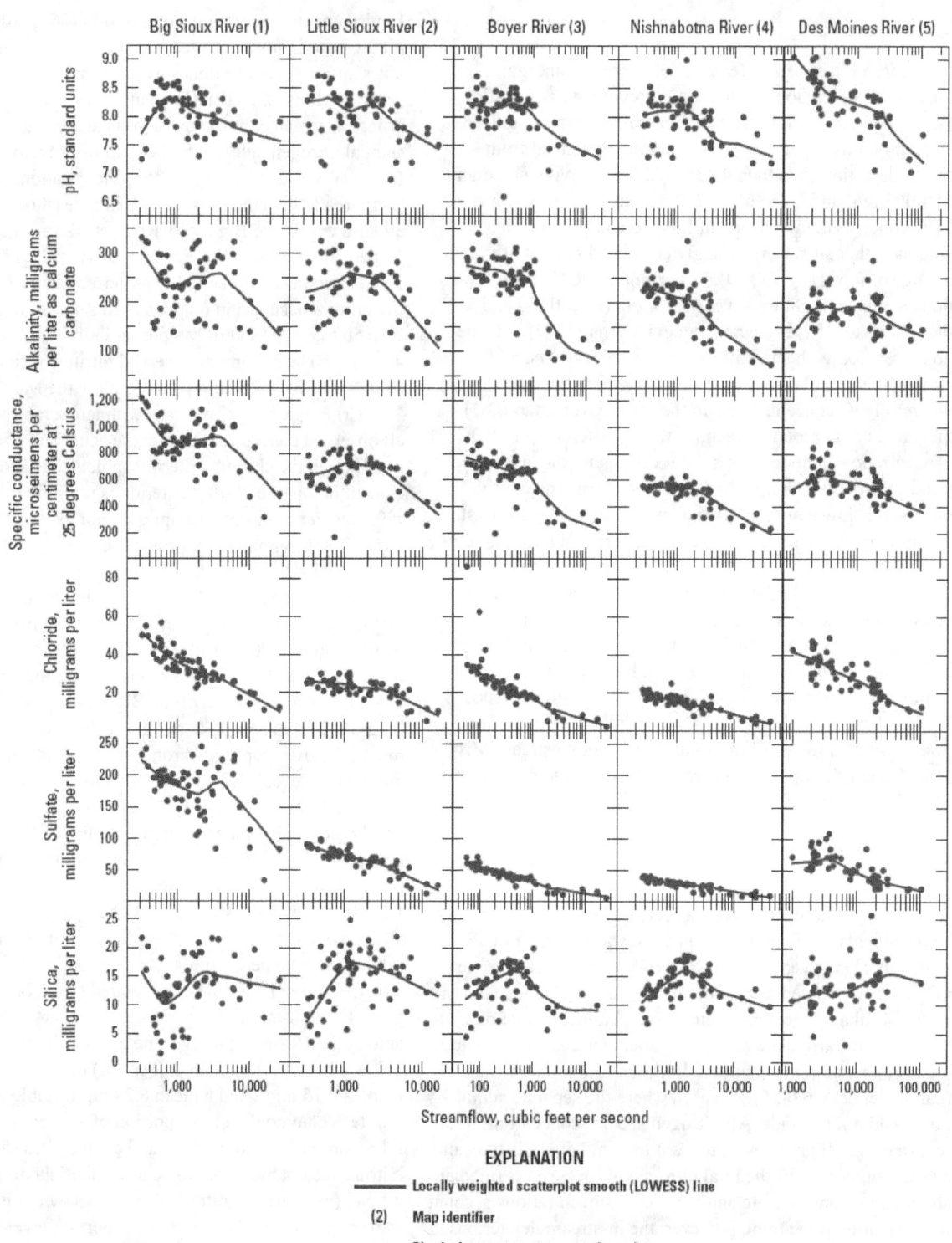

Figure 3A. Physical properties and concentrations related to streamflow for 10 major Iowa rivers with map identifiers and locally weighted scatterplot smooth (LOWESS) line, water years 2004–2008.

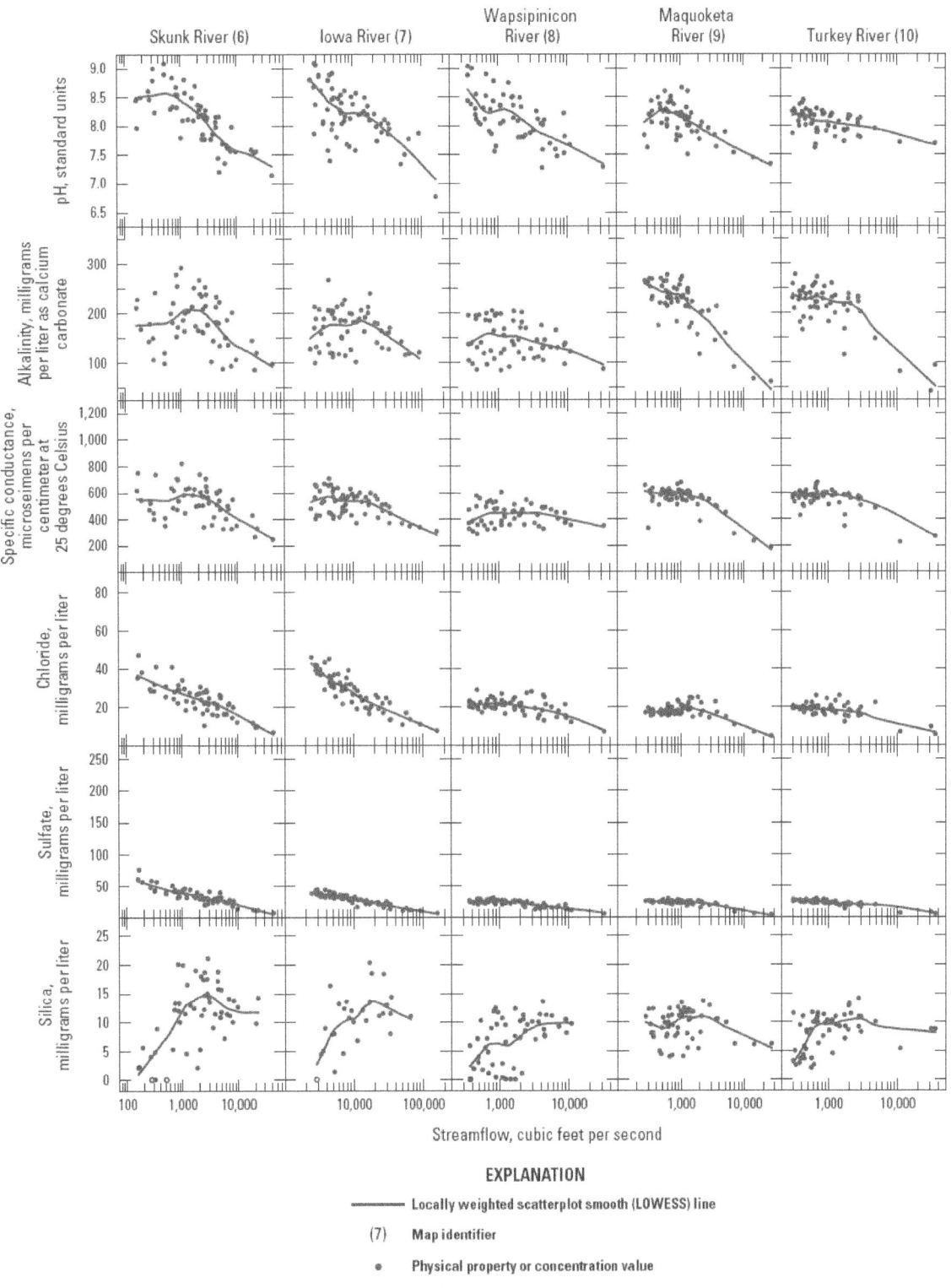

Figure 3*B*. Physical properties and concentrations related to streamflow for 10 major Iowa rivers with map identifiers and locally weighted scatterplot smooth (LOWESS) line, water years 2004–2008.—Continued

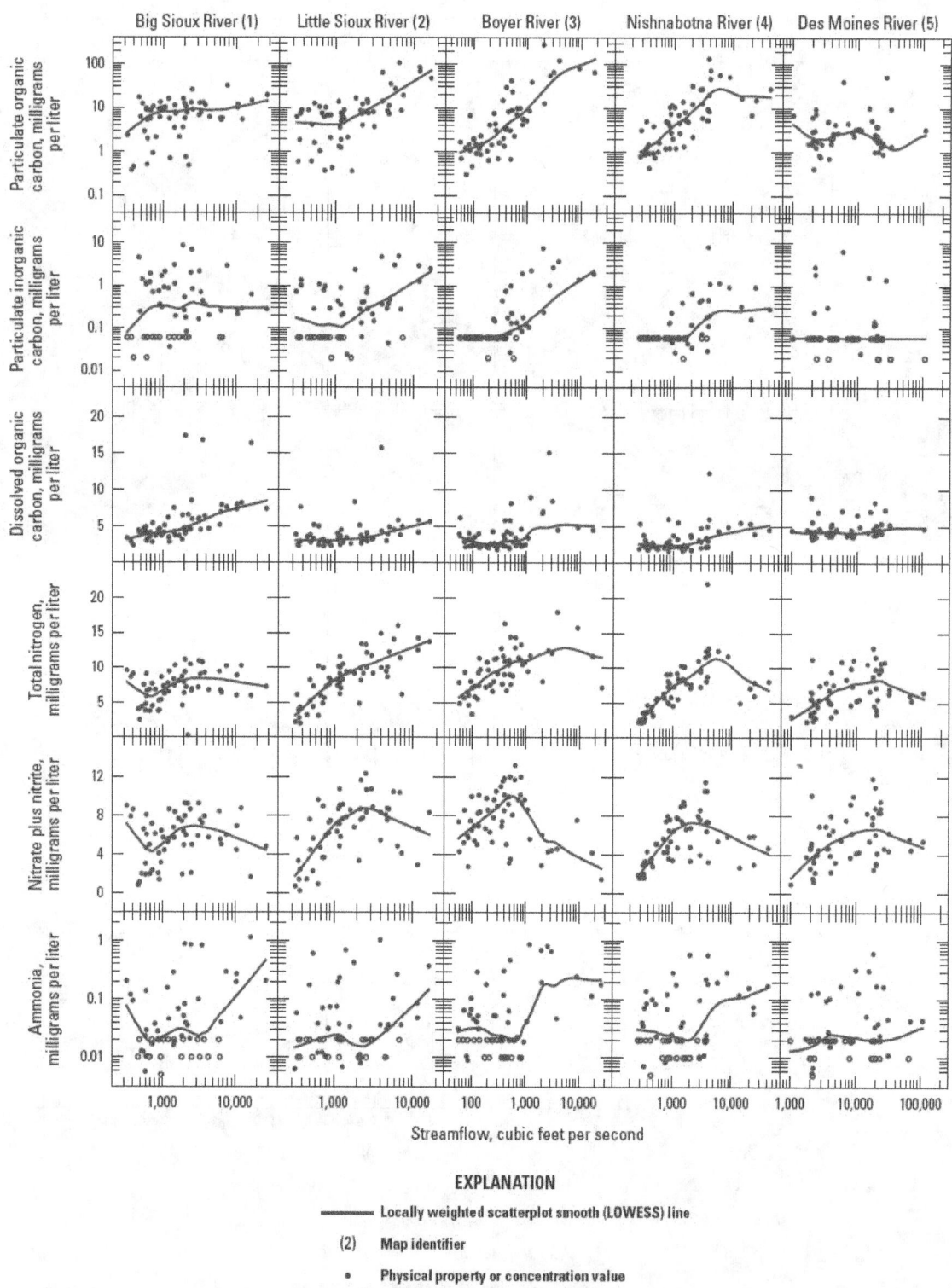

Figure 3C.   Physical properties and concentrations related to streamflow for 10 major Iowa rivers with map identifiers and locally weighted scatterplot smooth (LOWESS) line, water years 2004–2008.—Continued

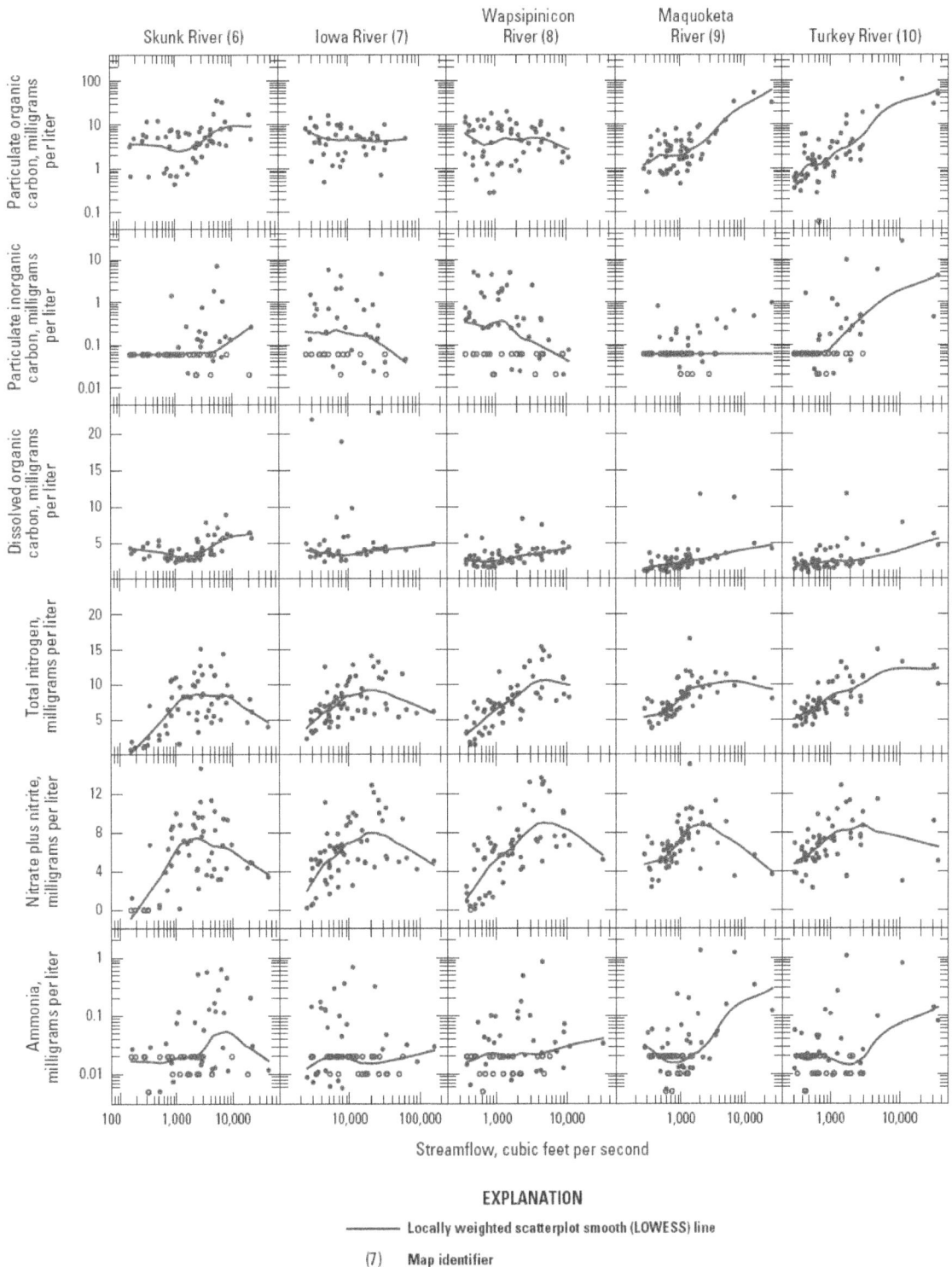

Figure 3*D*. Physical properties and concentrations related to streamflow for 10 major Iowa rivers with map identifiers and locally weighted scatterplot smooth (LOWESS) line, water years 2004–2008.—Continued

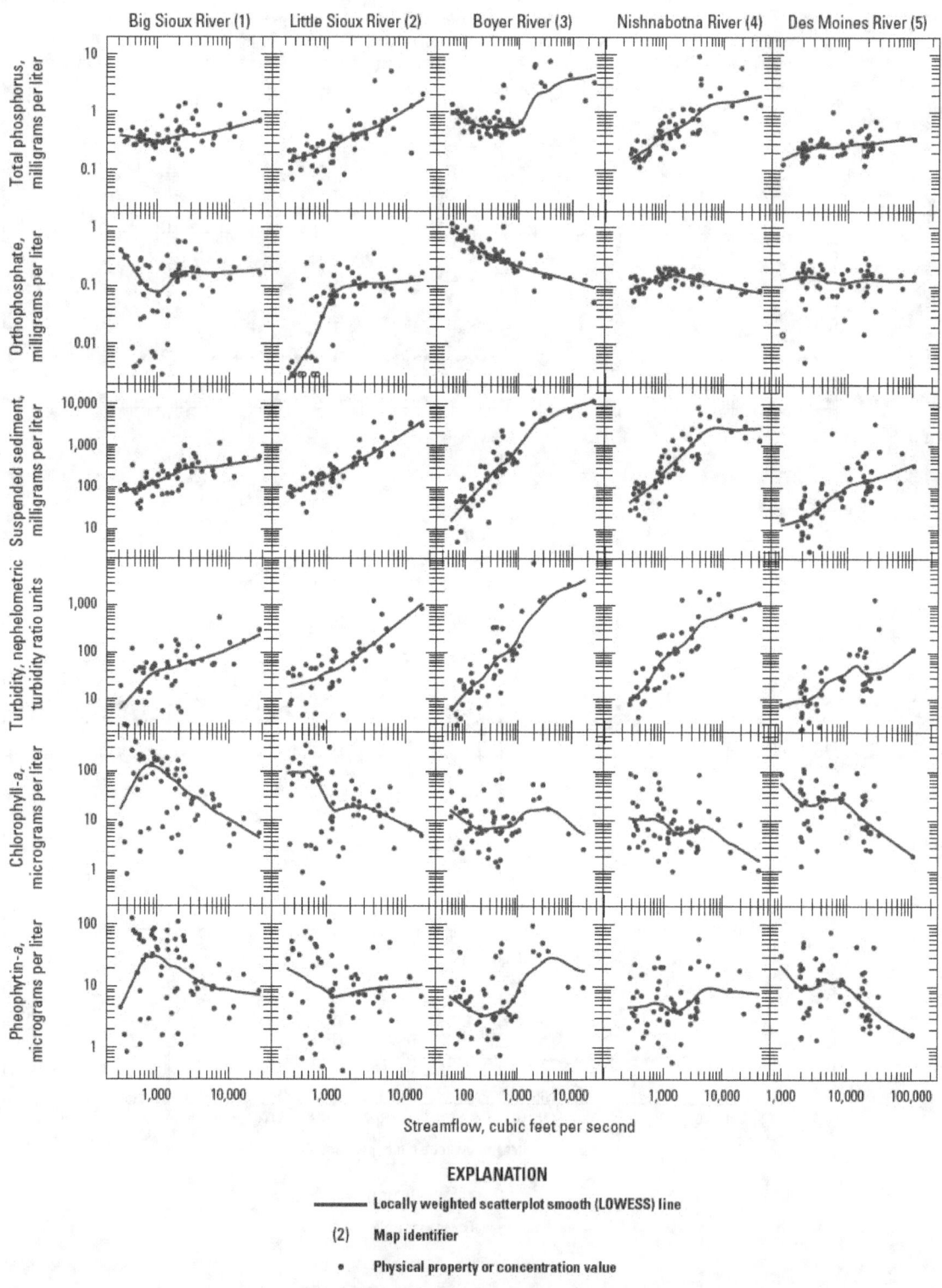

Figure 3E.   Physical properties and concentrations related to streamflow for 10 major Iowa rivers with map identifiers and locally weighted scatterplot smooth (LOWESS) line, water years 2004–2008.—Continued

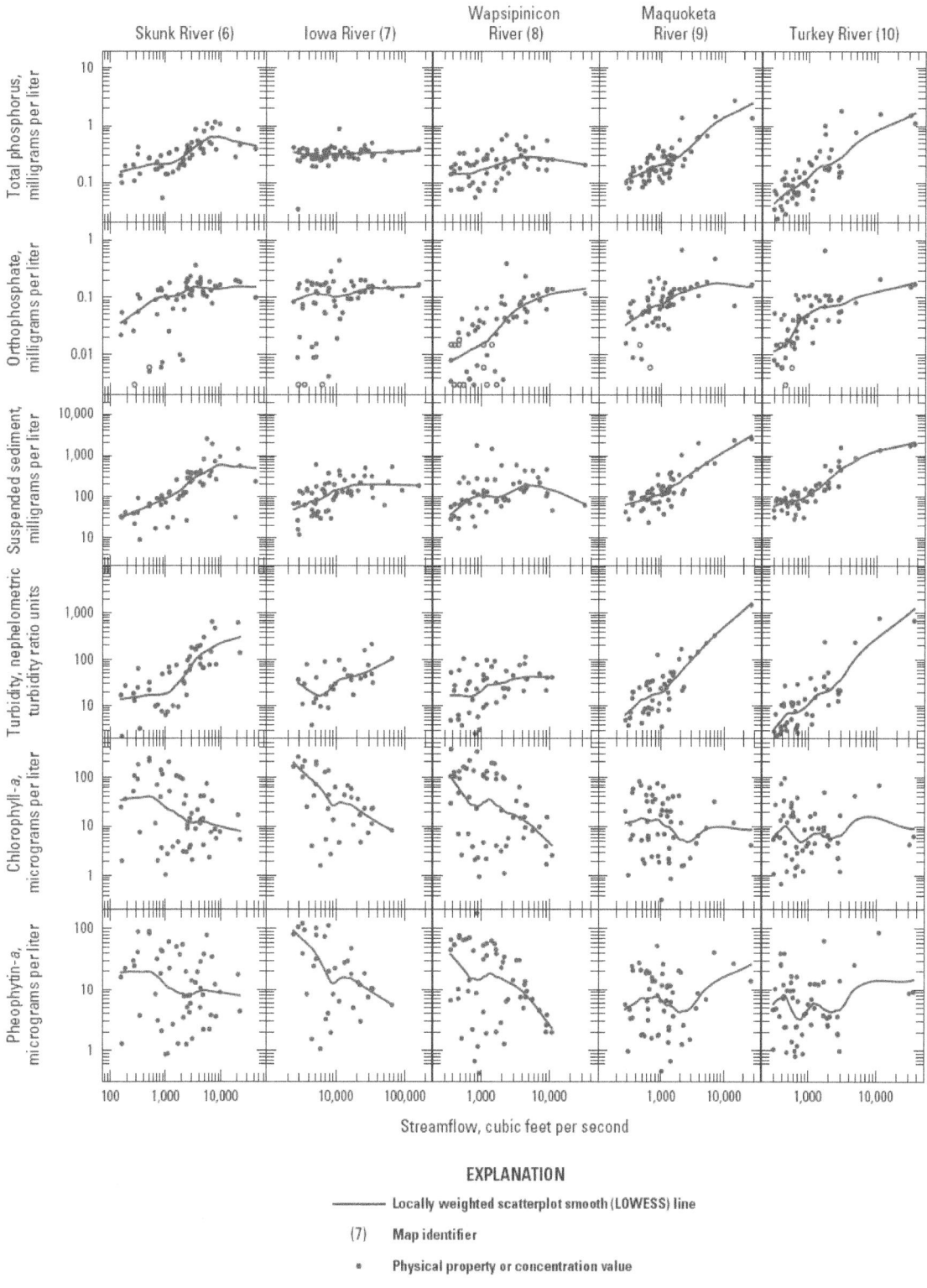

Figure 3F. Physical properties and concentrations related to streamflow for 10 major Iowa rivers with map identifiers and locally weighted scatterplot smooth (LOWESS) line, water years 2004–2008.—Continued

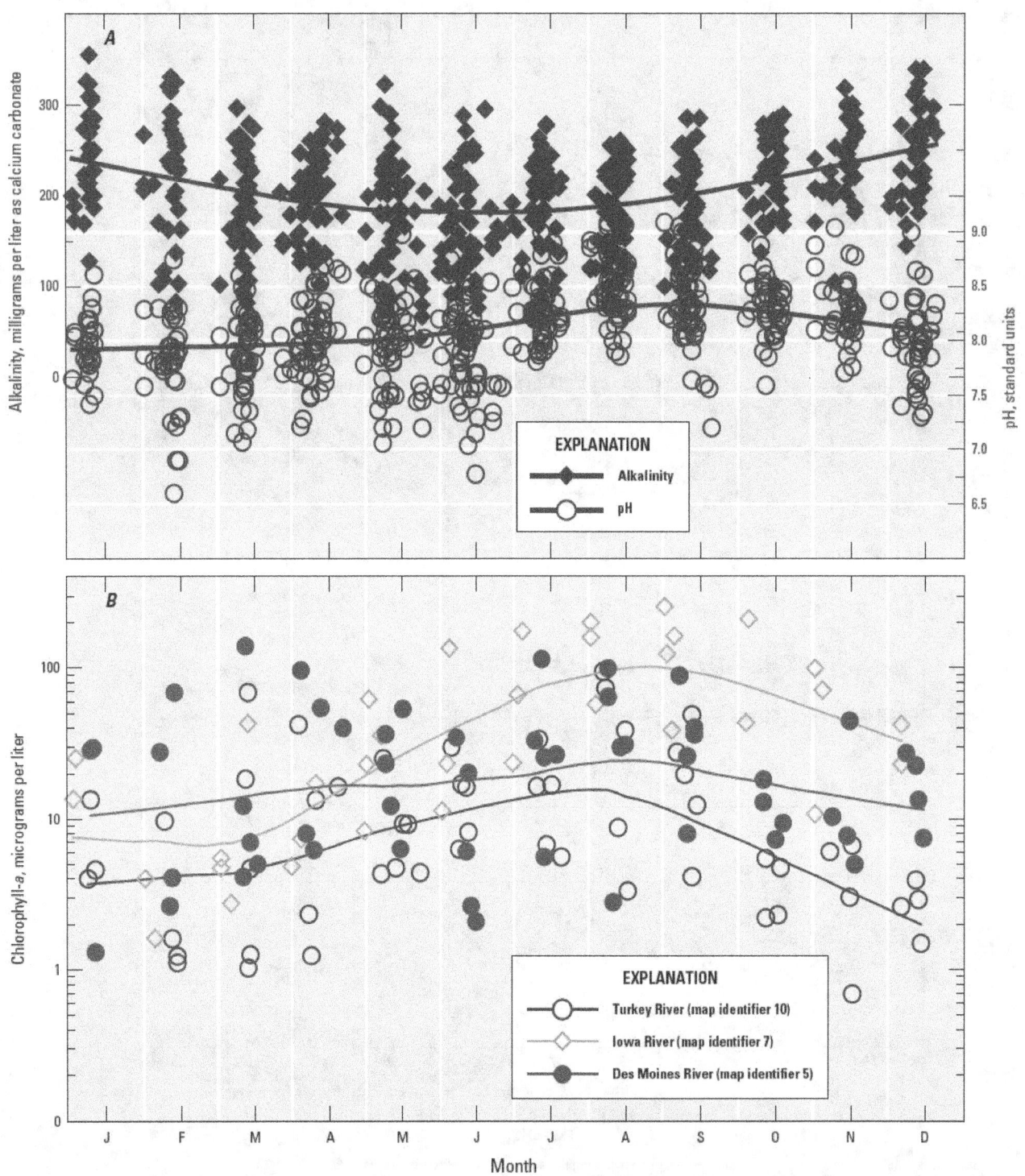

Figure 4. Seasonal values for *A*, Alkalinity and pH for 10 major Iowa rivers; *B*, chlorophyll-*a* concentrations; *C*, atrazine concentrations; and *D*, prometon concentrations for selected major Iowa rivers, water years 2004–2008 with locally weighted scatterplot smooth (LOWESS) line.

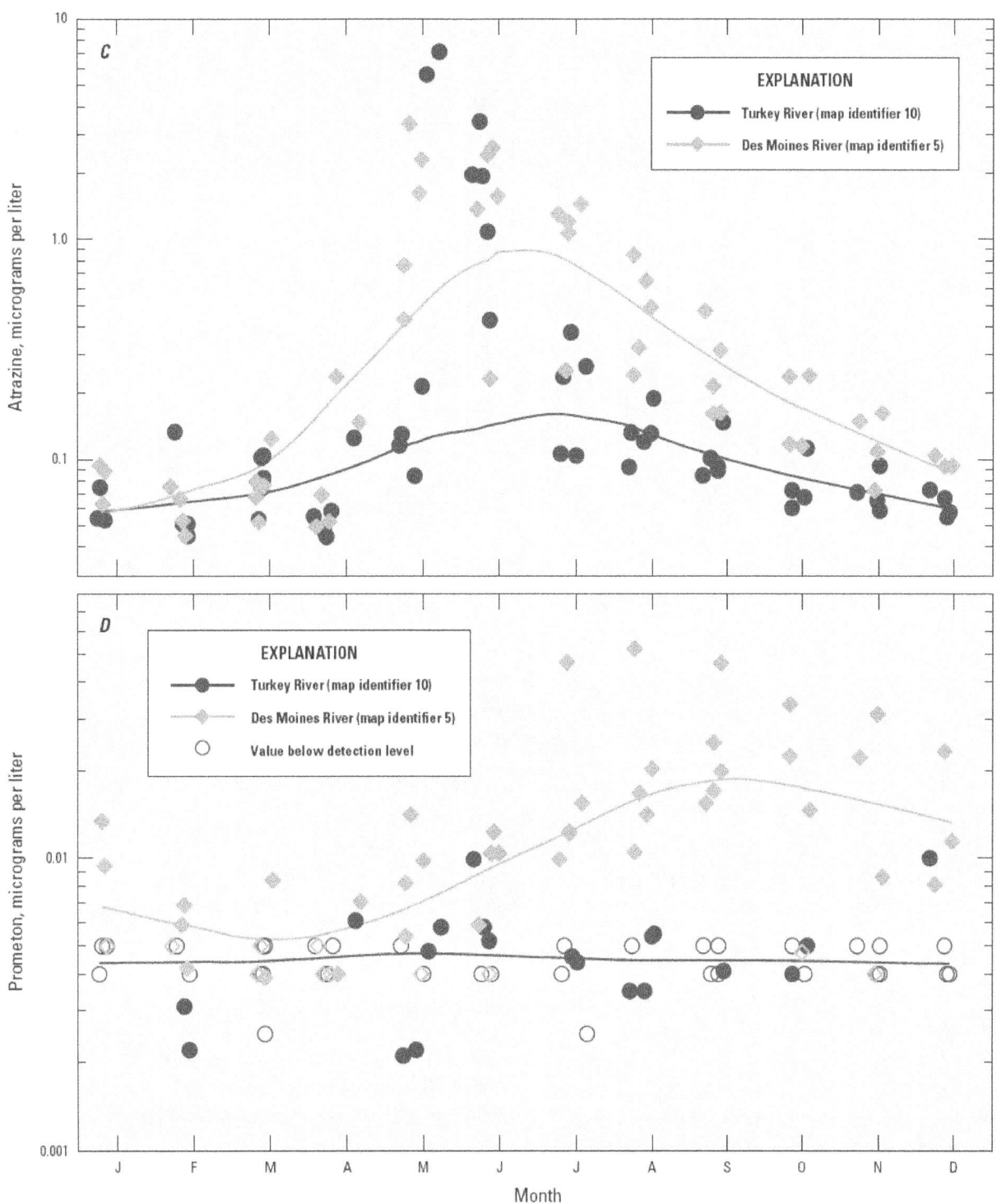

Figure 4. Seasonal values for *A*, Alkalinity and pH for 10 major Iowa rivers; *B*, chlorophyll-*a* concentrations; *C*, atrazine concentrations; and *D*, prometon concentrations for selected major Iowa rivers, water years 2004–2008 with locally weighted scatterplot smooth (LOWESS) line.—Continued

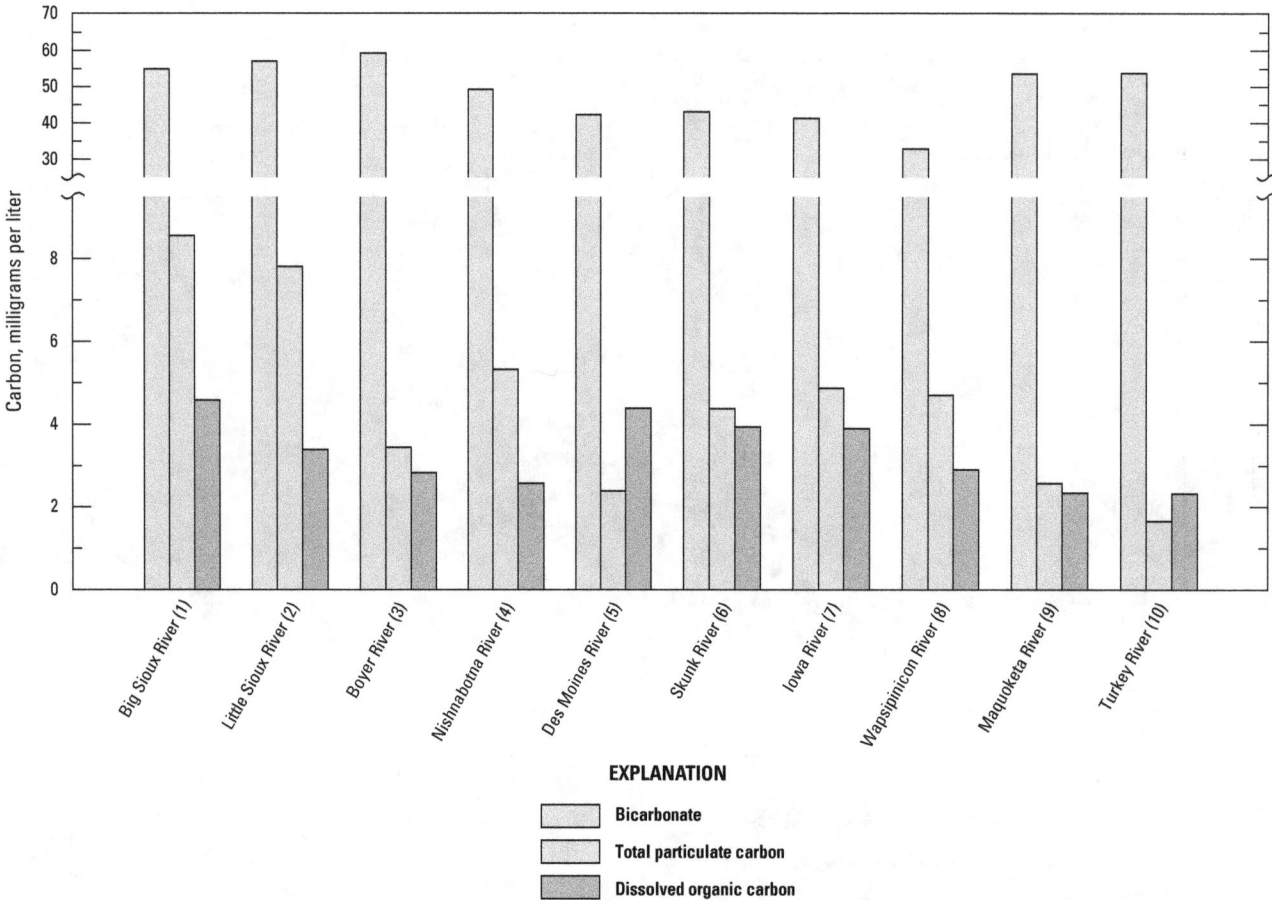

Figure 5.   Median carbon concentrations as bicarbonate, total particulate carbon, and dissolved organic carbon for 10 major Iowa rivers showing map identifiers, water years 2004–2008.

levels ranging from 0.01 to 0.04 mg/L (table 3). Sample concentrations for ammonia at high streamflows tended to be greater relative to low flows, with fewer results below detection (fig. 3C–D).

Phosphorus

Concentrations for phosphorus species are reported in milligrams per liter as phosphorus. Like nitrogen, phosphorus water-quality standards for the protection of aquatic life are still being developed in Iowa, though the Wisconsin total phosphorus (TP) criteria for large rivers is 0.1 mg/L (Wisconsin Department of Natural Resources, 2010). TP concentrations were equal to or greater than the Wisconsin criteria in 92 percent of samples, with four sites exceeding the criteria with every sample (Big Sioux, Boyer, Nishnabotna, and Des Moines Rivers; map IDs 1, 3, 4, and 5). TP concentrations ranging from 0.023 to 9.41 mg/L had a median 0.303 mg/L and a mean 0.507 mg/L (table 3). The correlation between TP concentrations and streamflow was generally positive, but varied across ranges of streamflow and among the sites (fig. 3E–F). TP concentrations were greatest and most variable

in the southwestern Iowa basins of the Boyer and Nishnabotna Rivers (map IDs 3–4), which are basins including steep regions of loess soils. TP concentrations indicated an overall downstream increase from Mississippi River tributaries. Orthophosphate concentrations ranged from below detection (reporting levels ranging from 0.006 to 0.036 mg/L) to 1.22 mg/L (table 3). The relation between orthophosphate concentrations and streamflow was generally flat to slightly positive, with the Boyer River (map ID 3) a notable exception having a strongly inverse relation to streamflow (fig. 3E–F).

Suspended Sediment and Turbidity

Suspended-sediment concentrations (SSC) ranged from 3 to 22,600 mg/L with a log-normal distribution around a median of 138 mg/L and a mean of 512 mg/L (table 3). SSC increased with streamflow, though the relation varied by site (fig. 3E–F). Sites in southwestern Iowa, Boyer and Nishnabotna Rivers, map IDs 3–4, had the greatest and most variable SSC; these basins include the highly erodible loess soils and steep slopes characteristic of the Western Loess Hills and Steeply Rolling Loess Prairie Ecoregions (fig. 2).

Turbidity values during sample collection ranged from 2.2 to 8,050 nephelometric turbidity ratio units (NTRU), with a median of 33 NTRU and a mean of 140 NTRU (table 3). Turbidity is related to suspended sediment, with a similar relation with streamflow among different sites, though the strength of the correlation is weaker at low turbidity and SSC values (figs. 3E–F, 6). Though the overall correlation R^2 is 0.68 between the log transform of these variables, sites with typically low turbidity and SSC values, such as the Wapsipinicon River (map ID 8), have a much weaker relation ($R^2 = 0.37$).

Algal Pigments

Algal pigments such as chlorophyll-*a* and pheophytin-*a* varied more by season than by streamflow, peaking in late summer (figs. 3E–F, 4B). Chlorophyll-*a* concentrations ranged from below detection (less than 0.1) to 384 μg/L with a log-normal distribution around a median 13.3 μg/L and a mean 38.5 μg/L (table 3). Pheophytin-*a* was similarly distributed, ranging from less than 0.1 to 181 μg/L with a median 8.3 μg/L and a mean 18.1 μg/L (table 3). The two algal pigments were strongly related to each other, with a linear correlation R^2 of 0.83 between the log transform of each variable.

Pesticides

Pesticide concentrations also varied by season, with greatest concentrations and most common detections occurring during typical application times. Two herbicides were found in every sample; atrazine and metolachlor, as well as the atrazine breakdown product 2-Chloro-4-isopropylamino-6-amino-s-triazine (CIAT; table 4). Atrazine concentrations exceeded the drinking-water MCL of 3.0 μg/L (U.S. Environmental Protection Agency, 2002) in 4 percent of the samples with the greatest concentration estimated at 41 μg/L, though none of the studied rivers are used directly for drinking-water sources near the sample collection sites. Atrazine and other commonly detected herbicides, such as acetochlor, alachlor, metribuzin, and simazine, had peak concentrations and most common detections from May through July (fig. 4C). Concentrations of prometon, a nonselective herbicide used frequently to control weeds in asphalt areas, were greatest from June through September, but occurrence varied among sites (fig. 4D). Insecticides were less commonly detected in samples; chlorpyrifos and fipronil were detected in 9 percent of all samples. Pesticides detected in water samples are summarized in table 4, and other pesticide compounds analyzed for but not detected during the study are listed in table 5.

Estimated Loads and Yields of Ions, Nutrients, and Suspended Sediment

The load and yield estimates described in this section for major ions (chloride, silica, and sulfate), nutrients (nitrate, total nitrogen, orthophosphate, and total phosphorus), and suspended sediment in 10 major Iowa tributaries to the Mississippi and Missouri Rivers include annual load estimates with standard error of prediction (SEP), upper and lower confidence limits, and basin yields (table 6). Models used for load estimates are described in table 7.

Streamflow and loads are positively correlated, even for sites and constituents with an inverse relation between flow and concentration. The year-to-year variations in loads (or yields) for any given site are largely because of streamflow variation. In general, largest calculated loads were for water years 2007 and 2008, which were the wettest years in the study period with the largest flows (table 1). Statewide monthly nutrient loads, for example, rise and fall with streamflow, peaking annually from April to June, with the greatest loads during June of 2008 when historic flooding occurred throughout much of the State (fig. 7, Buchmiller and Eash, 2010). Of the 10 sites, the Iowa River (map ID 7) and Des Moines River (map ID 5) Basins produced the largest loads for most constituents, which is exactly what would be expected given that their watershed areas are the two largest of all of the sites in the study, almost twice the watershed area of the next largest and more than an order of magnitude larger than the smallest watershed in the study.

The load estimates presented with this study tend to have narrower confidence limits than previously reported load estimates, where the same sites and constituents were analyzed, though results are generally comparable. Nitrate, orthophosphate, and total phosphorus loads presented by Aulenbach and others (2007) were derived from similar rating-curve equations in LOADEST for the Iowa River at Wapello (map ID 7). A different period of record was used for model calibration, and additional terms to describe streamflow variability were included for load estimates presented in this report. The differences in reported loads are not contradictory; rather, because the confidence limits overlap in all cases of comparable sites and constituents, the results are corroborated for both methods (fig. 8A–C). In the case of orthophosphate, confidence limits are not reported for estimates from this report because prediction intervals cannot be calculated using the LAD method, which was used where the assumption of normality of model residuals could not be met. Furthermore, the increased accuracy of estimates presented in this study demonstrates the usefulness of the streamflow variability terms. Overall, 38 of the individual models for constituents at each site incorporated streamflow variability terms (dQ_1, $|dQ_1|$, or dQ_{30}), and an additional six models used streamflow anomalies (A5yr, A1yr, A3mo, HFV). Because model selection procedures emphasized low residual variance (good model fit), the inclusion of these terms indicates improvements in the accuracy of the predictions, as well.

The model residuals for suspended sediment tended to be greater than other constituents, because of the high variability in SSC. The SEP for annual suspended-sediment loads for each site and year ranged from 10 to 49 percent (table 6), with an average of 21 percent. Among the other constituents, only orthophosphate and total phosphorus average annual SEP

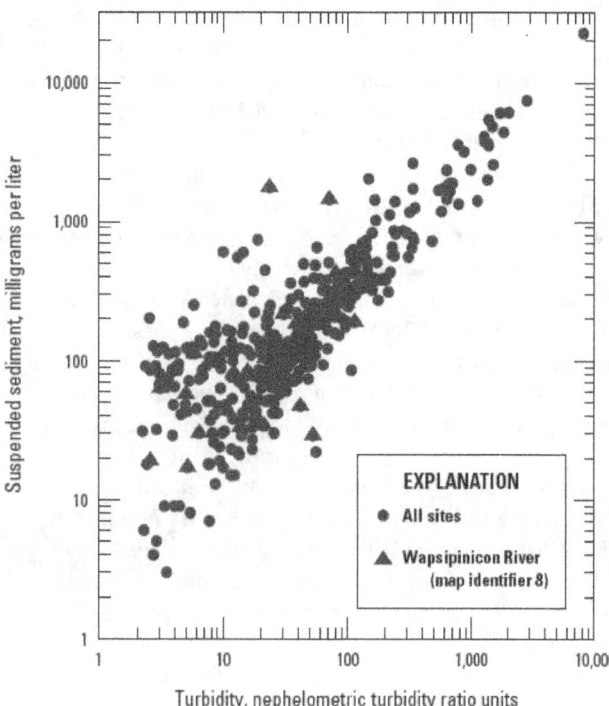

Figure 6.   Relations between suspended-sediment concentrations and turbidity for 10 major Iowa rivers, water years 2004–2008.

exceeded 10 percent (table 6). One site with a daily suspended-sediment record was used to validate the suspended-sediment loads estimated by LOADEST. Load estimates are published annually (Nalley and others, 2005a, b; U.S. Geological Survey, 2007–2009) for the Skunk River at Augusta (map ID 6) based on suspended sediment and streamflow time-series daily observations and using the USGS Graphical Constituent Loading Analysis System (GCLAS; Koltun and others, 2006). The SEP for annual suspended-sediment load by rating curve method (LOADEST) from the Skunk River ranged from 19 to 32 percent (table 6). The relative percent difference (RPD, calculated as the difference divided by the mean) between annual LOADEST and GCLAS estimates ranged from 4 to 32 percent (fig. 8D). Predictive errors cannot be calculated for GCLAS load estimates because it is not a regression technique, and instead fits a smoothed time series for load through every sampled concentration.

Constituent yields tended to be greater during wet years, but similarities and geographic patterns among sites varied by constituent (fig. 9). Yields related to annual streamflow scaled to the long-term average streamflow (1979–2008; table 1) for each site are shown on figure 9. This representation of the x-axis allows for easy comparison of wet (greater than 1) and dry (less than 1) years among sites. Constituents with similar patterns in yields among sites suggest that transport in the basins is governed by similar processes. Iowa's Missouri River tributaries consistently yielded less chloride and more

sediment than Mississippi River tributaries. The Big Sioux River Basin yields were distinct from other sites for many constituents, including sulfate, silica, total nitrogen, nitrate, total phosphorus, and suspended sediment.

Major Ions

Annual chloride loads ranged from 3,670 tons from the Boyer River (map ID 3) in 2006 to 348,000 tons from the Iowa River (map ID 7) in 2008, with SEP not exceeding 5 percent (table 6). Smaller basins generated greater loads for the same streamflows as larger basins. The lowest annual yield was 3.29 tons per square mile (ton/mi^2) from the Nishnabotna River (map ID 4) in 2006 and the greatest was 28.6 ton/mi^2 from the Wapsipinicon River (map ID 8) in 2008. Missouri River tributaries tended to have lower yields than Mississippi River tributaries (fig. 9A).

Annual sulfate loads ranged from 6,850 tons from the Boyer River (map ID 3) in 2006 to 612,000 tons from the Des Moines River (map ID 5) in 2008 (table 6). Annual sulfate yields ranged from 6.09 ton/mi^2 from the Nishnabotna River (map ID 4) to 57.6 tons/mi^2 from the Big Sioux River (map ID 1), both occurring in 2006. The Big Sioux, Little Sioux (map ID 2), and Des Moines Rivers, had the greatest sulfate loads and yields, and the yields compared to long term streamflow for these sites indicated different patterns than the other seven sites (fig. 9B).

Silica loads were smallest in 2006 at the Boyer River (map ID 3) at 1,830 tons and greatest in 2008 at the Des Moines River (map ID 5) at 312,000 tons (table 6). The Des Moines and Iowa (map ID 7) Rivers had the greatest annual silica loads by a factor of 2 to 3 above any other site, because of the greater streamflow in these rivers. The smallest and largest silica yields were estimated at 1.84 ton/mi^2 from the Big Sioux (map ID 1) in 2004 and 24.2 ton/mi^2 from the Iowa River in 2008. The Big Sioux River had the lowest silica yield over the 5-year study period, and yields at this site had a distinct pattern with streamflow compared to other sites (fig. 9C).

Nitrogen

TN loads ranged from 1,200 tons in 2006 from the Boyer River (map ID 3) to 145,000 tons in 2008 from the Iowa River (map ID 7; table 6). The Boyer River had the smallest TN loads whereas the Des Moines (map ID 5) and Iowa Rivers consistently had the largest loads of the study sites, reflective of basin size and overall streamflow. The smallest annual yield was estimated at 1.09 ton/mi^2 in 2006 at the Nishnabotna River (map ID 4), and the largest yield was 19.3 ton/mi^2 in 2008 at the Maquoketa River (map ID 9). On average, the Big Sioux River (map ID 1) had the smallest TN yield during the study period (less than one-half the yield of any other river), and northeastern basins (map IDs 7–10) tended to have larger yields than western and central basins (fig. 9D).

Table 6. Estimated major ions, nitrogen, phosphorus, and suspended-sediment loads and yields from selected major Iowa rivers, water years 2004–2008.

[Water year from October 1 to September 30; ID, identifier; SEP, standard error of prediction; L95 and U95, lower (L) and upper (U) limits of the 95-percent confidence interval of estimated load; ton/mi², tons per square mile; mg/L, milligrams per liter; parameter code given in brackets, see also table 2; µg/L, micrograms per liter; --, not available]

Map ID (fig. 2)	Station name	2004					
		Load	SEP		L95	U95	Yield
		(tons)	(tons)	(percent)	(tons)	(tons)	(ton/mi²)
	Chloride, in mg/L [00940]						
1	Big Sioux River at Akron, Iowa	38,300	1,170	3.1	36,000	40,600	5.47
2	Little Sioux River at Turin, Iowa	24,700	686	2.8	23,400	26,100	6.95
3	Boyer River at Logan, Iowa	6,280	150	2.4	5,990	6,580	7.21
4	Nishnabotna River above Hamburg, Iowa	17,500	453	2.6	16,700	18,400	6.24
5	Des Moines River at Keosauqua, Iowa	194,000	8,980	4.6	177,000	212,000	13.8
6	Skunk River at Augusta, Iowa	50,200	1,260	2.5	47,800	52,700	11.6
7	Iowa River at Wapello, Iowa	261,000	6,770	2.6	248,000	275,000	20.9
8	Wapsipinicon River near De Witt, Iowa	38,500	1,360	3.5	35,900	41,200	16.5
9	Maquoketa River near Spragueville, Iowa	23,500	510	2.2	22,600	24,600	14.4
10	Turkey River at Garber, Iowa	22,100	730	3.3	20,700	23,600	14.2
	Total	**677,000**	**22,100**	**3.3**	**634,000**	**721,000**	**13.4**
	Sulfate, in mg/L [00945]						
1	Big Sioux River at Akron, Iowa	185,000	7,960	4.3	169,000	201,000	26.4
2	Little Sioux River at Turin, Iowa	71,000	1,600	2.3	67,900	74,200	20.0
3	Boyer River at Logan, Iowa	11,900	331	2.8	11,300	12,600	13.7
4	Nishnabotna River above Hamburg, Iowa	33,200	781	2.4	31,700	34,800	11.8
5	Des Moines River at Keosauqua, Iowa	352,000	13,400	3.8	327,000	379,000	25.1
6	Skunk River at Augusta, Iowa	69,500	1,620	2.3	66,400	72,800	16.1
7	Iowa River at Wapello, Iowa	243,000	3,980	1.6	235,000	251,000	19.4
8	Wapsipinicon River near De Witt, Iowa	41,900	1,000	2.4	40,000	43,900	17.9
9	Maquoketa River near Spragueville, Iowa	29,300	529	1.8	28,300	30,400	18.0
10	Turkey River at Garber, Iowa	27,600	474	1.7	26,600	28,500	17.7
	Total	**1,060,000**	**31,700**	**3.0**	**1,000,000**	**1,130,000**	**21.0**
	Silica, in mg/L [00955]						
1	Big Sioux River at Akron, Iowa	12,900	--	--	--	--	1.84
2	Little Sioux River at Turin, Iowa	14,700	1,470	10	12,000	17,800	4.14
3	Boyer River at Logan, Iowa	4,700	225	4.8	4,270	5,150	5.40
4	Nishnabotna River above Hamburg, Iowa	17,700	--	--	--	--	6.29
5	Des Moines River at Keosauqua, Iowa	98,000	4,420	4.5	89,600	107,000	6.98
6	Skunk River at Augusta, Iowa	31,500	--	--	--	--	7.29
7	Iowa River at Wapello, Iowa	109,000	--	--	--	--	8.72
8	Wapsipinicon River near De Witt, Iowa	20,100	--	--	--	--	8.59
9	Maquoketa River near Spragueville, Iowa	12,000	1,220	10	9,830	14,600	7.37
10	Turkey River at Garber, Iowa	12,100	1,800	15	9,000	16,000	7.82
	Total	**333,000**	**--**	**--**	**--**	**--**	**6.59**

Table 6. Estimated major ions, nitrogen, phosphorus, and suspended-sediment loads and yields from selected major Iowa rivers, water years 2004–2008.—Continued

[Water year from October 1 to September 30; ID, identifier; SEP, standard error of prediction; L95 and U95, lower (L) and upper (U) limits of the 95-percent confidence interval of estimated load; ton/mi², tons per square mile; mg/L, milligrams per liter; parameter code given in brackets, see also table 2; μg/L, micrograms per liter; --, not available]

Map ID (fig. 2)	Station name	2004					
		Load	SEP		L95	U95	Yield
		(tons)	(tons)	(percent)	(tons)	(tons)	(ton/mi²)
Total nitrogen, in mg/L [49570 plus 62854, or 62855]							
1	Big Sioux River at Akron, Iowa	9,040	422	4.7	8,240	9,890	1.29
2	Little Sioux River at Turin, Iowa	12,400	586	4.7	11,200	13,500	3.48
3	Boyer River at Logan, Iowa	3,630	203	5.6	3,250	4,050	4.18
4	Nishnabotna River above Hamburg, Iowa	12,500	569	4.6	11,400	13,600	4.44
5	Des Moines River at Keosauqua, Iowa	51,800	4,580	8.8	43,400	61,300	3.69
6	Skunk River at Augusta, Iowa	22,000	1,780	8.1	18,800	25,700	5.11
7	Iowa River at Wapello, Iowa	79,700	19,000	24	48,900	123,000	6.38
8	Wapsipinicon River near De Witt, Iowa	19,300	2,000	10	15,700	23,500	8.27
9	Maquoketa River near Spragueville, Iowa	12,500	473	3.8	11,600	13,400	7.65
10	Turkey River at Garber, Iowa	14,800	955	6.5	13,000	16,800	9.53
	Total	**238,000**	**30,600**	**13**	**185,000**	**305,000**	**4.71**
Nitrate plus nitrite, in mg/L [00631]							
1	Big Sioux River at Akron, Iowa	6,490	574	8.8	5,440	7,690	0.928
2	Little Sioux River at Turin, Iowa	7,850	569	7.2	6,800	9,030	2.21
3	Boyer River at Logan, Iowa	2,660	147	5.5	2,390	2,960.0	3.06
4	Nishnabotna River above Hamburg, Iowa	10,200	848	8.3	8,680	12,000	3.65
5	Des Moines River at Keosauqua, Iowa	43,300	4,780	11	34,700	53,400	3.08
6	Skunk River at Augusta, Iowa	16,100	1,350	8.4	13,600	18,900	3.74
7	Iowa River at Wapello, Iowa	68,500	8,690	13	53,100	87,100	5.48
8	Wapsipinicon River near De Witt, Iowa	17,200	2,140	12	13,400	21,800	7.38
9	Maquoketa River near Spragueville, Iowa	9,750	415	4.3	8,960	10,600	5.97
10	Turkey River at Garber, Iowa	10,200	624	6.1	8,980	11,400	6.53
	Total	**192,000**	**20,100**	**10**	**156,000**	**235,000**	**3.80**
Total phosphorus, in mg/L [00665]							
1	Big Sioux River at Akron, Iowa	683	53.1	7.8	585	793	0.0976
2	Little Sioux River at Turin, Iowa	1,430	267	19	981	2,020	0.404
3	Boyer River at Logan, Iowa	406	69.0	17	287	558	0.466
4	Nishnabotna River above Hamburg, Iowa	2,020	272	13	1,540	2,610	0.721
5	Des Moines River at Keosauqua, Iowa	2,210	100	4.5	2,020	2,410	0.157
6	Skunk River at Augusta, Iowa	1,360	132	9.7	1,120	1,640	0.315
7	Iowa River at Wapello, Iowa	2,870	167	5.8	2,560	3,210	0.230
8	Wapsipinicon River near De Witt, Iowa	649	85.4	13	498	832	0.278
9	Maquoketa River near Spragueville, Iowa	675	74.4	11	541	832	0.413
10	Turkey River at Garber, Iowa	1,500	549	37	706	2,830	0.968
	Total	**13,800**	**1,770**	**13**	**10,800**	**17,700**	**0.273**

Table 6. Estimated major ions, nitrogen, phosphorus, and suspended-sediment loads and yields from selected major Iowa rivers, water years 2004–2008.—Continued

[Water year from October 1 to September 30; ID, identifier; SEP, standard error of prediction; L95 and U95, lower (L) and upper (U) limits of the 95-percent confidence interval of estimated load; ton/mi², tons per square mile; mg/L, milligrams per liter; parameter code given in brackets, see also table 2; µg/L, micrograms per liter; --, not available]

Map ID (fig. 2)	Station name	2004					
		Load	SEP		L95	U95	Yield
		(tons)	(tons)	(percent)	(tons)	(tons)	(ton/mi²)
		Orthophosphate, in mg/L [00671]					
1	Big Sioux River at Akron, Iowa	304	166	55	98.0	729	0.0435
2	Little Sioux River at Turin, Iowa	158	39.8	25	93.9	249	0.0444
3	Boyer River at Logan, Iowa	113	5.51	4.9	102	124	0.129
4	Nishnabotna River above Hamburg, Iowa	192	8.95	4.7	175	210	0.0683
5	Des Moines River at Keosauqua, Iowa	808	--	--	--	--	0.0576
6	Skunk River at Augusta, Iowa	331	--	--	--	--	0.0766
7	Iowa River at Wapello, Iowa	1,020	--	--	--	--	0.0812
8	Wapsipinicon River near De Witt, Iowa	241	109	45	93.7	513	0.103
9	Maquoketa River near Spragueville, Iowa	202	35.6	18	141	280	0.123
10	Turkey River at Garber, Iowa	190	62.3	33	96.3	337	0.122
	Total	**3,560**	--	--	--	--	**0.0705**
		Suspended sediment, in mg/L [80154]					
1	Big Sioux River at Akron, Iowa	427,000	61,400	14	319,000	559,000	61.0
2	Little Sioux River at Turin, Iowa	2,060,000	401,000	19	1,390,000	2,950,000	580
3	Boyer River at Logan, Iowa	587,000	168,000	29	326,000	977,000	674
4	Nishnabotna River above Hamburg, Iowa	3,620,000	898,000	25	2,180,000	5,670,000	1,290
5	Des Moines River at Keosauqua, Iowa	1,800,000	425,000	24	1,110,000	2,770,000	128
6	Skunk River at Augusta, Iowa	1,380,000	342,000	25	832,000	2,160,000	320
7	Iowa River at Wapello, Iowa	2,300,000	395,000	17	1,620,000	3,170,000	184
8	Wapsipinicon River near De Witt, Iowa	405,000	68,300	17	287,000	554,000	173
9	Maquoketa River near Spragueville, Iowa	661,000	141,000	21	428,000	978,000	405
10	Turkey River at Garber, Iowa	1,350,000	424,000	31	704,000	2,350,000	867
	Total	**14,600,000**	**3,320,000**	**23**	**9,190,000**	**22,100,000**	**289**

Table 6. Estimated major ions, nitrogen, phosphorus, and suspended-sediment loads and yields from selected major Iowa rivers, water years 2004–2008.—Continued

[Water year from October 1 to September 30; ID, identifier; SEP, standard error of prediction; L95 and U95, lower (L) and upper (U) limits of the 95-percent confidence interval of estimated load; ton/mi^2, tons per square mile; mg/L, milligrams per liter; parameter code given in brackets, see also table 2; μg/L, micrograms per liter; --, not available]

Map ID (fig. 2)	Station name	2005					
		Load	SEP		L95	U95	Yield
		(tons)	(tons)	(percent)	(tons)	(tons)	(ton/mi^2)
	Chloride, in mg/L [00940]						
1	Big Sioux River at Akron, Iowa	44,500	796	1.8	43,000	46,100	6.36
2	Little Sioux River at Turin, Iowa	36,200	949	2.6	34,400	38,100	10.2
3	Boyer River at Logan, Iowa	5,630	132	2.3	5,370	5,890	6.47
4	Nishnabotna River above Hamburg, Iowa	15,100	380	2.5	14,400	15,800	5.37
5	Des Moines River at Keosauqua, Iowa	159,000	6,740	4.2	146,000	172,000	11.3
6	Skunk River at Augusta, Iowa	46,400	1,190	2.6	44,100	48,800	10.8
7	Iowa River at Wapello, Iowa	199,000	3,480	1.7	192,000	206,000	15.9
8	Wapsipinicon River near De Witt, Iowa	24,300	476	2.0	23,400	25,300	10.4
9	Maquoketa River near Spragueville, Iowa	12,300	261	2.1	11,800	12,800	7.52
10	Turkey River at Garber, Iowa	13,700	340	2.5	13,100	14,400	8.85
	Total	**556,000**	**14,800**	**2.7**	**528,000**	**586,000**	**11.0**
	Sulfate, in mg/L [00945]						
1	Big Sioux River at Akron, Iowa	235,000	7,390	3.1	221,000	250,000	33.6
2	Little Sioux River at Turin, Iowa	103,000	2,210	2.1	98,700	107,000	29.0
3	Boyer River at Logan, Iowa	10,600	292	2.8	10,100	11,200	12.2
4	Nishnabotna River above Hamburg, Iowa	28,300	647	2.3	27,000	29,600	10.1
5	Des Moines River at Keosauqua, Iowa	331,000	12,200	3.7	308,000	356,000	23.6
6	Skunk River at Augusta, Iowa	64,700	1,550	2.4	61,800	67,800	15.0
7	Iowa River at Wapello, Iowa	220,000	3,620	1.6	213,000	227,000	17.6
8	Wapsipinicon River near De Witt, Iowa	26,000	544	2.1	24,900	27,100	11.1
9	Maquoketa River near Spragueville, Iowa	16,500	281	1.7	16,000	17,100	10.1
10	Turkey River at Garber, Iowa	17,900	229	1.3	17,500	18,400	11.5
	Total	**1,050,000**	**29,000**	**2.8**	**998,000**	**1,110,000**	**20.8**
	Silica, in mg/L [00955]						
1	Big Sioux River at Akron, Iowa	16,600	--	--	--	--	2.37
2	Little Sioux River at Turin, Iowa	25,700	1,810	7.0	22,400	29,500	7.25
3	Boyer River at Logan, Iowa	3,870	177	4.6	3,540	4,230	4.45
4	Nishnabotna River above Hamburg, Iowa	13,800	--	--	--	--	4.93
5	Des Moines River at Keosauqua, Iowa	80,300	3,400	4.2	73,900	87,200	5.73
6	Skunk River at Augusta, Iowa	26,000	--	--	--	--	6.04
7	Iowa River at Wapello, Iowa	73,200	--	--	--	--	5.85
8	Wapsipinicon River near De Witt, Iowa	6,370	--	--	--	--	2.73
9	Maquoketa River near Spragueville, Iowa	5,750	330	5.7	5,130	6,420	3.52
10	Turkey River at Garber, Iowa	5,630	481	8.5	4,750	6,630	3.62
	Total	**257,000**	**--**	**--**	**--**	**--**	**5.09**

Table 6.   Estimated major ions, nitrogen, phosphorus, and suspended-sediment loads and yields from selected major Iowa rivers, water years 2004–2008.—Continued

[Water year from October 1 to September 30; ID, identifier; SEP, standard error of prediction; L95 and U95, lower (L) and upper (U) limits of the 95-percent confidence interval of estimated load; ton/mi², tons per square mile; mg/L, milligrams per liter; parameter code given in brackets, see also table 2; μg/L, micrograms per liter; --, not available]

Map ID (fig. 2)	Station name	2005					
		Load	SEP		L95	U95	Yield
		(tons)	(tons)	(percent)	(tons)	(tons)	(ton/mi²)
	Total nitrogen, in mg/L [49570 plus 62854, or 62855]						
1	Big Sioux River at Akron, Iowa	10,700	468	4.4	9,770	11,600	1.53
2	Little Sioux River at Turin, Iowa	16,600	716	4.3	15,300	18,100	4.68
3	Boyer River at Logan, Iowa	2,870	156	5.4	2,580	3,190	3.30
4	Nishnabotna River above Hamburg, Iowa	8,860	398	4.5	8,100	9,660	3.15
5	Des Moines River at Keosauqua, Iowa	58,700	3,750	6.4	51,700	66,400	4.18
6	Skunk River at Augusta, Iowa	18,700	1,500	8.0	15,900	21,800	4.33
7	Iowa River at Wapello, Iowa	81,800	12,800	16	59,500	110,000	6.54
8	Wapsipinicon River near De Witt, Iowa	7,950	493	6.2	7,030	8,960	3.41
9	Maquoketa River near Spragueville, Iowa	5,180	187	3.6	4,830	5,560	3.17
10	Turkey River at Garber, Iowa	5,680	202	3.6	5,290	6,080	3.65
	Total	**217,000**	**20,700**	**10**	**180,000**	**261,000**	**4.29**
	Nitrate plus nitrite, in mg/L [00631]						
1	Big Sioux River at Akron, Iowa	7,850	625	8.0	6,700	9,150	1.12
2	Little Sioux River at Turin, Iowa	13,900	891	6.4	12,200	15,700	3.91
3	Boyer River at Logan, Iowa	2,200	114	5.2	1,980	2,430	2.53
4	Nishnabotna River above Hamburg, Iowa	6,740	395	5.9	6,000	7,550	2.40
5	Des Moines River at Keosauqua, Iowa	49,800	4,040	8.1	42,300	58,200	3.55
6	Skunk River at Augusta, Iowa	14,500	1,170	8.1	12,300	16,900	3.36
7	Iowa River at Wapello, Iowa	56,000	4,150	7.4	48,300	64,500	4.48
8	Wapsipinicon River near De Witt, Iowa	6,970	733	11	5,650	8,520	2.98
9	Maquoketa River near Spragueville, Iowa	4,200	179	4.3	3,860	4,570	2.57
10	Turkey River at Garber, Iowa	4,520	194	4.3	4,150	4,910	2.91
	Total	**167,000**	**12,500**	**7.5**	**144,000**	**192,000**	**3.30**
	Total phosphorus, in mg/L [00665]						
1	Big Sioux River at Akron, Iowa	671	37.1	5.5	601	747	0.0959
2	Little Sioux River at Turin, Iowa	944	110	12	748	1,180	0.266
3	Boyer River at Logan, Iowa	380	46.5	12	297	479	0.436
4	Nishnabotna River above Hamburg, Iowa	1160	135	12	920	1,450	0.414
5	Des Moines River at Keosauqua, Iowa	1920	87.0	4.5	1,750	2,090	0.137
6	Skunk River at Augusta, Iowa	912	75.4	8.3	773	1,070	0.211
7	Iowa River at Wapello, Iowa	2,200	78.3	3.6	2,050	2,360	0.176
8	Wapsipinicon River near De Witt, Iowa	239	22.9	10	197	286	0.102
9	Maquoketa River near Spragueville, Iowa	171	14.8	8.7	144	202	0.105
10	Turkey River at Garber, Iowa	169	28.2	17	120	230	0.109
	Total	**8,770**	**635**	**7.2**	**7,600**	**10,100**	**0.173**

Table 6.   Estimated major ions, nitrogen, phosphorus, and suspended-sediment loads and yields from selected major Iowa rivers, water years 2004–2008.—Continued

[Water year from October 1 to September 30; ID, identifier; SEP, standard error of prediction; L95 and U95, lower (L) and upper (U) limits of the 95-percent confidence interval of estimated load; ton/mi², tons per square mile; mg/L, milligrams per liter; parameter code given in brackets, see also table 2; µg/L, micrograms per liter; --, not available]

Map ID (fig. 2)	Station name	2005					
		Load	SEP		L95	U95	Yield
		(tons)	(tons)	(percent)	(tons)	(tons)	(ton/mi²)
	Orthophosphate, in mg/L [00671]						
1	Big Sioux River at Akron, Iowa	291	80.3	28	165	477	0.0416
2	Little Sioux River at Turin, Iowa	258	58.1	23	163	390	0.0727
3	Boyer River at Logan, Iowa	96.0	6.88	7.2	83.3	110	0.110
4	Nishnabotna River above Hamburg, Iowa	148	6.66	4.5	135	161	0.0526
5	Des Moines River at Keosauqua, Iowa	737	--	--	--	--	0.0525
6	Skunk River at Augusta, Iowa	233	--	--	--	--	0.0539
7	Iowa River at Wapello, Iowa	813	--	--	--	--	0.0650
8	Wapsipinicon River near De Witt, Iowa	49.0	14.3	29	26.9	82.4	0.0210
9	Maquoketa River near Spragueville, Iowa	69.6	15.0	22	44.8	103	0.0426
10	Turkey River at Garber, Iowa	51.3	12.2	24	31.5	79.0	0.0330
	Total	**2,750**	--	--	--	--	**0.0544**
	Suspended sediment, in mg/L [80154]						
1	Big Sioux River at Akron, Iowa	425,000	48,200	11	339,000	527,000	60.8
2	Little Sioux River at Turin, Iowa	1,100,000	123,000	11	875,000	1,360,000	308
3	Boyer River at Logan, Iowa	443,000	172,000	39	199,000	860,000	510
4	Nishnabotna River above Hamburg, Iowa	1,920,000	466,000	24	1,160,000	2,980,000	682
5	Des Moines River at Keosauqua, Iowa	1,330,000	324,000	24	810,000	2,070,000	95.1
6	Skunk River at Augusta, Iowa	744,000	141,000	19	506,000	1,060,000	172
7	Iowa River at Wapello, Iowa	1,330,000	193,000	15	995,000	1,750,000	107
8	Wapsipinicon River near De Witt, Iowa	157,000	22,900	15	117,000	207,000	67.2
9	Maquoketa River near Spragueville, Iowa	222,000	56,900	26	131,000	353,000	136
10	Turkey River at Garber, Iowa	129,000	15,600	12	101,000	162,000	83.0
	Total	**7,800,000**	**1,560,000**	**20**	**5,240,000**	**11,300,000**	**154**

Table 6. Estimated major ions, nitrogen, phosphorus, and suspended-sediment loads and yields from selected major Iowa rivers, water years 2004–2008.—Continued

[Water year from October 1 to September 30; ID, identifier; SEP, standard error of prediction; L95 and U95, lower (L) and upper (U) limits of the 95-percent confidence interval of estimated load; ton/mi², tons per square mile; mg/L, milligrams per liter; parameter code given in brackets, see also table 2; µg/L, micrograms per liter; --, not available]

Map ID (fig. 2)	Station name	2006					
		Load	SEP		L95	U95	Yield
		(tons)	(tons)	(percent)	(tons)	(tons)	(ton/mi²)
	Chloride, in mg/L [00940]						
1	Big Sioux River at Akron, Iowa	65,300	1,400	2.1	62,600	68,100	9.33
2	Little Sioux River at Turin, Iowa	34,600	949	2.7	32,800	36,500	9.75
3	Boyer River at Logan, Iowa	3,670	103	2.8	3,470	3,870	4.22
4	Nishnabotna River above Hamburg, Iowa	9,250	276	3.0	8,720	9,800	3.29
5	Des Moines River at Keosauqua, Iowa	161,000	5,430	3.4	151,000	172,000	11.5
6	Skunk River at Augusta, Iowa	27,000	757	2.8	25,500	28,500	6.25
7	Iowa River at Wapello, Iowa	191,000	3,750	2.0	184,000	199,000	15.3
8	Wapsipinicon River near De Witt, Iowa	30,100	685	2.3	28,700	31,400	12.9
9	Maquoketa River near Spragueville, Iowa	12,700	273	2.1	12,200	13,300	7.78
10	Turkey River at Garber, Iowa	16,100	410	2.5	15,300	16,900	10.3
	Total	**551,000**	**14,000**	**2.5**	**524,000**	**579,000**	**10.9**
	Sulfate, in mg/L [00945]						
1	Big Sioux River at Akron, Iowa	403,000	13,600	3.4	377,000	430,000	57.6
2	Little Sioux River at Turin, Iowa	98,200	2,220	2.3	93,900	103,000	27.6
3	Boyer River at Logan, Iowa	6,850	216	3.2	6,430	7,280	7.87
4	Nishnabotna River above Hamburg, Iowa	17,100	505	3.0	16,100	18,100	6.09
5	Des Moines River at Keosauqua, Iowa	298,000	11,800	4.0	276,000	322,000	21.3
6	Skunk River at Augusta, Iowa	38,400	1,000	2.6	36,400	40,400	8.89
7	Iowa River at Wapello, Iowa	190,000	3,180	1.7	184,000	197,000	15.2
8	Wapsipinicon River near De Witt, Iowa	29,800	618	2.1	28,600	31,100	12.8
9	Maquoketa River near Spragueville, Iowa	17,200	287	1.7	16,600	17,700	10.5
10	Turkey River at Garber, Iowa	20,700	272	1.3	20,100	21,200	13.3
	Total	**1,120,000**	**33,600**	**3.0**	**1,050,000**	**1,190,000**	**22.2**
	Silica, in mg/L [00955]						
1	Big Sioux River at Akron, Iowa	34,300	--	--	--	--	4.90
2	Little Sioux River at Turin, Iowa	27,600	2,240	8.1	23,400	32,200	7.76
3	Boyer River at Logan, Iowa	1,830	79.5	4.3	1,680	1,990	2.10
4	Nishnabotna River above Hamburg, Iowa	6,650	--	--	--	--	2.37
5	Des Moines River at Keosauqua, Iowa	59,000	2,460	4.2	54,300	64,000	4.21
6	Skunk River at Augusta, Iowa	12,000	--	--	--	--	2.78
7	Iowa River at Wapello, Iowa	58,900	--	--	--	--	4.71
8	Wapsipinicon River near De Witt, Iowa	7,940	--	--	--	--	3.40
9	Maquoketa River near Spragueville, Iowa	5,560	343	6.2	4,920	6,260	3.40
10	Turkey River at Garber, Iowa	7,790	587	7.5	6,710	9,000	5.01
	Total	**222,000**	**--**	**--**	**--**	**--**	**4.39**

Table 6. Estimated major ions, nitrogen, phosphorus, and suspended-sediment loads and yields from selected major Iowa rivers, water years 2004–2008.—Continued

[Water year from October 1 to September 30; ID, identifier; SEP, standard error of prediction; L95 and U95, lower (L) and upper (U) limits of the 95-percent confidence interval of estimated load; ton/mi², tons per square mile; mg/L, milligrams per liter; parameter code given in brackets, see also table 2; μg/L, micrograms per liter; --, not available]

Map ID (fig. 2)	Station name	2006					
		Load	SEP		L95	U95	Yield
		(tons)	(tons)	(percent)	(tons)	(tons)	(ton/mi²)
	Total nitrogen, in mg/L [49570 plus 62854, or 62855]						
1	Big Sioux River at Akron, Iowa	20,200	1,020	5.0	18,300	22,200	2.89
2	Little Sioux River at Turin, Iowa	17,600	856	4.9	16,000	19,300	4.95
3	Boyer River at Logan, Iowa	1,200	62.2	5.2	1,090	1,330	1.38
4	Nishnabotna River above Hamburg, Iowa	3,060	129	4.2	2,810	3,320	1.09
5	Des Moines River at Keosauqua, Iowa	43,400	2,750	6.3	38,300	49,100	3.10
6	Skunk River at Augusta, Iowa	8,930	820	9.2	7,430	10,600	2.07
7	Iowa River at Wapello, Iowa	69,000	10,200	15	51,200	91,100	5.52
8	Wapsipinicon River near De Witt, Iowa	10,600	665	6.3	9,350	12,000	4.54
9	Maquoketa River near Spragueville, Iowa	5,050	171	3.4	4,720	5,390	3.09
10	Turkey River at Garber, Iowa	7,280	282	3.9	6,740	7,850	4.69
	Total	**186,000**	**17,000**	**9.1**	**156,000**	**222,000**	**3.68**
	Nitrate plus nitrite, in mg/L [00631]						
1	Big Sioux River at Akron, Iowa	18,700	1,900	10	15,200	22,700	2.67
2	Little Sioux River at Turin, Iowa	14,300	1,060	7.4	12,400	16,500	4.03
3	Boyer River at Logan, Iowa	1,040	52.6	5.1	945	1,150	1.20
4	Nishnabotna River above Hamburg, Iowa	2,030	127	6.3	1,790	2,290	0.723
5	Des Moines River at Keosauqua, Iowa	35,400	2,870	8.1	30,100	41,400	2.52
6	Skunk River at Augusta, Iowa	6,680	590	8.8	5,600	7,910	1.55
7	Iowa River at Wapello, Iowa	44,200	3,410	7.7	37,900	51,200	3.53
8	Wapsipinicon River near De Witt, Iowa	9,490	995	10	7,690	11,600	4.06
9	Maquoketa River near Spragueville, Iowa	4,030	181	4.5	3,680	4,390	2.46
10	Turkey River at Garber, Iowa	6,220	278	4.5	5,690	6,780	4.00
	Total	**142,000**	**11,500**	**8.1**	**121,000**	**166,000**	**2.81**
	Total phosphorus, in mg/L [00665]						
1	Big Sioux River at Akron, Iowa	978	70.5	7.2	847	1,120	0.140
2	Little Sioux River at Turin, Iowa	984	114	12	779	1,230	0.277
3	Boyer River at Logan, Iowa	129	13.4	10	105	158	0.148
4	Nishnabotna River above Hamburg, Iowa	295	30.1	10	240	358	0.105
5	Des Moines River at Keosauqua, Iowa	1,500	68.9	4.6	1,370	1,640	0.107
6	Skunk River at Augusta, Iowa	486	51.9	11	392	595	0.113
7	Iowa River at Wapello, Iowa	1,930	61.3	3.2	1,810	2,050	0.154
8	Wapsipinicon River near De Witt, Iowa	312	30.5	10	256	376	0.133
9	Maquoketa River near Spragueville, Iowa	174	11.0	6.3	154	197	0.107
10	Turkey River at Garber, Iowa	214	30.7	14	160	280	0.138
	Total	**7,000**	**482**	**6.9**	**6,110**	**8,000**	**0.138**

Table 6. Estimated major ions, nitrogen, phosphorus, and suspended-sediment loads and yields from selected major Iowa rivers, water years 2004–2008.—Continued

[Water year from October 1 to September 30; ID, identifier; SEP, standard error of prediction; L95 and U95, lower (L) and upper (U) limits of the 95-percent confidence interval of estimated load; ton/mi², tons per square mile; mg/L, milligrams per liter; parameter code given in brackets, see also table 2; μg/L, micrograms per liter; --, not available]

Map ID (fig. 2)	Station name	2006					
		Load	SEP		L95	U95	Yield
		(tons)	(tons)	(percent)	(tons)	(tons)	(ton/mi²)
Orthophosphate, in mg/L [00671]							
1	Big Sioux River at Akron, Iowa	689	244	35	331	1,270	0.0985
2	Little Sioux River at Turin, Iowa	319	85.7	27	184	517	0.0899
3	Boyer River at Logan, Iowa	85.1	6.65	7.8	72.8	98.9	0.0978
4	Nishnabotna River above Hamburg, Iowa	72.9	3.42	4.7	66.5	79.9	0.0260
5	Des Moines River at Keosauqua, Iowa	601	--	--	--	--	0.0428
6	Skunk River at Augusta, Iowa	102	--	--	--	--	0.0235
7	Iowa River at Wapello, Iowa	607	--	--	--	--	0.0486
8	Wapsipinicon River near De Witt, Iowa	63.5	15.0	24	39.2	97.5	0.0272
9	Maquoketa River near Spragueville, Iowa	52.4	7.49	14	39.2	68.5	0.0321
10	Turkey River at Garber, Iowa	73.8	16.0	22	47.4	110	0.0475
	Total	**2,670**	--	--	--	--	**0.0528**
Suspended sediment, in mg/L [80154]							
1	Big Sioux River at Akron, Iowa	614,000	76,700	12	478,000	778,000	87.8
2	Little Sioux River at Turin, Iowa	1,160,000	143,000	12	910,000	1,470,000	328
3	Boyer River at Logan, Iowa	53,000	13,500	25	31,400	84,100	60.9
4	Nishnabotna River above Hamburg, Iowa	306,000	66,800	22	196,000	457,000	109
5	Des Moines River at Keosauqua, Iowa	727,000	171,000	24	449,000	1,120,000	51.8
6	Skunk River at Augusta, Iowa	385,000	97,500	25	229,000	608,000	89.2
7	Iowa River at Wapello, Iowa	949,000	124,000	13	729,000	1,220,000	75.9
8	Wapsipinicon River near De Witt, Iowa	211,000	31,300	15	156,000	278,000	90.2
9	Maquoketa River near Spragueville, Iowa	86,100	12,800	15	63,700	114,000	52.7
10	Turkey River at Garber, Iowa	154,000	15,600	10	126,000	187,000	99.1
	Total	**4,650,000**	**752,000**	**16**	**3,370,000**	**6,300,000**	**92.0**

Table 6.   Estimated major ions, nitrogen, phosphorus, and suspended-sediment loads and yields from selected major Iowa rivers, water years 2004–2008.—Continued

[Water year from October 1 to September 30; ID, identifier; SEP, standard error of prediction; L95 and U95, lower (L) and upper (U) limits of the 95-percent confidence interval of estimated load; ton/mi^2, tons per square mile; mg/L, milligrams per liter; parameter code given in brackets, see also table 2; μg/L, micrograms per liter; --, not available]

Map ID (fig. 2)	Station name	2007					
		Load	SEP		L95	U95	Yield
		(tons)	(tons)	(percent)	(tons)	(tons)	(ton/mi²)
	Chloride, in mg/L [00940]						
1	Big Sioux River at Akron, Iowa	57,900	1,230	2.1	55,500	60,400	8.28
2	Little Sioux River at Turin, Iowa	40,500	1,330	3.3	38,000	43,200	11.4
3	Boyer River at Logan, Iowa	10,600	301	2.8	10,000	11,200	12.2
4	Nishnabotna River above Hamburg, Iowa	29,500	791	2.7	28,000	31,100	10.5
5	Des Moines River at Keosauqua, Iowa	283,000	10,500	3.7	264,000	305,000	20.2
6	Skunk River at Augusta, Iowa	76,800	2,150	2.8	72,700	81,100	17.8
7	Iowa River at Wapello, Iowa	305,000	6,490	2.1	293,000	318,000	24.4
8	Wapsipinicon River near De Witt, Iowa	52,500	1,190	2.3	50,200	54,900	22.5
9	Maquoketa River near Spragueville, Iowa	22,500	486	2.2	21,600	23,500	13.8
10	Turkey River at Garber, Iowa	25,400	802	3.2	23,800	27,000	16.3
	Total	**904,000**	**25,200**	**2.8**	**856,000**	**955,000**	**17.9**
	Sulfate, in mg/L [00945]						
1	Big Sioux River at Akron, Iowa	378,000	14,700	3.9	350,000	408,000	54.0
2	Little Sioux River at Turin, Iowa	114,000	2,870	2.5	109,000	120,000	32.2
3	Boyer River at Logan, Iowa	20,400	649	3.2	19,200	21,700	23.4
4	Nishnabotna River above Hamburg, Iowa	57,900	1,660	2.9	54,700	61,200	20.6
5	Des Moines River at Keosauqua, Iowa	503,000	18,600	3.7	468,000	541,000	35.9
6	Skunk River at Augusta, Iowa	105,000	2,730	2.6	99,900	111,000	24.4
7	Iowa River at Wapello, Iowa	318,000	5,760	1.8	306,000	329,000	25.4
8	Wapsipinicon River near De Witt, Iowa	51,600	1,200	2.3	49,300	54,000	22.1
9	Maquoketa River near Spragueville, Iowa	28,300	509	1.8	27,300	29,300	17.3
10	Turkey River at Garber, Iowa	31,500	522	1.7	30,500	32,500	20.3
	Total	**1,610,000**	**49,200**	**3.1**	**1,510,000**	**1,710,000**	**31.8**
	Silica, in mg/L [00955]						
1	Big Sioux River at Akron, Iowa	34,300	--	--	--	--	4.90
2	Little Sioux River at Turin, Iowa	37,300	3,090	8.3	31,600	43,700	10.5
3	Boyer River at Logan, Iowa	9,810	535	5.5	8,800	10,900	11.3
4	Nishnabotna River above Hamburg, Iowa	38,500	--	--	--	--	13.7
5	Des Moines River at Keosauqua, Iowa	183,000	8,440	4.6	167,000	201,000	13.1
6	Skunk River at Augusta, Iowa	53,800	--	--	--	--	12.5
7	Iowa River at Wapello, Iowa	150,000	--	--	--	--	12.0
8	Wapsipinicon River near De Witt, Iowa	24,200	--	--	--	--	10.4
9	Maquoketa River near Spragueville, Iowa	11,300	683	6.0	10,100	12,700	6.94
10	Turkey River at Garber, Iowa	17,300	1,700	10	14,200	20,900	11.2
	Total	**561,000**	--	--	--	--	**11.1**

Table 6. Estimated major ions, nitrogen, phosphorus, and suspended-sediment loads and yields from selected major Iowa rivers, water years 2004–2008.—Continued

[Water year from October 1 to September 30; ID, identifier; SEP, standard error of prediction; L95 and U95, lower (L) and upper (U) limits of the 95-percent confidence interval of estimated load; ton/mi², tons per square mile; mg/L, milligrams per liter; parameter code given in brackets, see also table 2; μg/L, micrograms per liter; --, not available]

Map ID (fig. 2)	Station name	2007					
		Load	SEP		L95	U95	Yield
		(tons)	(tons)	(percent)	(tons)	(tons)	(ton/mi²)
Total nitrogen, in mg/L [49570 plus 62854, or 62855]							
1	Big Sioux River at Akron, Iowa	18,300	1,040	5.7	16,400	20,400	2.62
2	Little Sioux River at Turin, Iowa	25,100	1,460	5.8	22,400	28,100	7.07
3	Boyer River at Logan, Iowa	8,420	523	6.2	7,440	9,490	9.68
4	Nishnabotna River above Hamburg, Iowa	24,800	1,150	4.6	22,600	27,100	8.82
5	Des Moines River at Keosauqua, Iowa	114,000	7,270	6.4	99,900	128,000	8.09
6	Skunk River at Augusta, Iowa	36,600	3,230	8.8	30,700	43,300	8.48
7	Iowa River at Wapello, Iowa	124,000	31,500	25	73,800	196,000	9.93
8	Wapsipinicon River near De Witt, Iowa	25,200	1,830	7.3	21,800	28,900	10.8
9	Maquoketa River near Spragueville, Iowa	11,400	413	3.6	10,600	12,200	6.95
10	Turkey River at Garber, Iowa	15,200	728	4.8	13,800	16,600	9.75
	Total	**403,000**	**49,200**	**12**	**319,000**	**511,000**	**7.97**
Nitrate plus nitrite, in mg/L [00631]							
1	Big Sioux River at Akron, Iowa	16,700	1,970	12	13,100	20,900	2.39
2	Little Sioux River at Turin, Iowa	19,700	1,800	9.1	16,400	23,500	5.55
3	Boyer River at Logan, Iowa	5,550	352	6.3	4,890	6,270	6.37
4	Nishnabotna River above Hamburg, Iowa	19,400	1,270	6.5	17,000	22,000	6.90
5	Des Moines River at Keosauqua, Iowa	101,000	8,320	8.2	86,000	119,000	7.22
6	Skunk River at Augusta, Iowa	27,500	2,470	9.0	22,900	32,600	6.36
7	Iowa River at Wapello, Iowa	104,000	8,250	7.9	88,900	121,000	8.33
8	Wapsipinicon River near De Witt, Iowa	24,400	2,700	11	19,600	30,100	10.5
9	Maquoketa River near Spragueville, Iowa	8,990	381	4.2	8,260	9,760	5.50
10	Turkey River at Garber, Iowa	12,500	745	6.0	11,100	14,000	8.05
	Total	**340,000**	**28,300**	**8.3**	**288,000**	**399,000**	**6.72**
Total phosphorus, in mg/L [00665]							
1	Big Sioux River at Akron, Iowa	985	78.3	7.9	841	1,150	0.141
2	Little Sioux River at Turin, Iowa	1,790	288	16	1,300	2,420	0.505
3	Boyer River at Logan, Iowa	1,290	267	21	846	1,890	1.48
4	Nishnabotna River above Hamburg, Iowa	2,520	279	11	2,020	3,110	0.897
5	Des Moines River at Keosauqua, Iowa	3,980	184	4.6	3,630	4,350	0.283
6	Skunk River at Augusta, Iowa	2,320	227	9.8	1,910	2,790	0.537
7	Iowa River at Wapello, Iowa	4,090	128	3.1	3850	4,350	0.328
8	Wapsipinicon River near De Witt, Iowa	706	73.7	10	573	861	0.302
9	Maquoketa River near Spragueville, Iowa	473	35.2	7.4	408	545	0.289
10	Turkey River at Garber, Iowa	645	113	18	453	893	0.415
	Total	**18,800**	**1,670**	**8.9**	**15,800**	**22,400**	**0.372**

Table 6. Estimated major ions, nitrogen, phosphorus, and suspended-sediment loads and yields from selected major Iowa rivers, water years 2004–2008.—Continued

[Water year from October 1 to September 30; ID, identifier; SEP, standard error of prediction; L95 and U95, lower (L) and upper (U) limits of the 95-percent confidence interval of estimated load; ton/mi², tons per square mile; mg/L, milligrams per liter; parameter code given in brackets, see also table 2; µg/L, micrograms per liter; --, not available]

Map ID (fig. 2)	Station name	2007					
		Load	SEP		L95	U95	Yield
		(tons)	(tons)	(percent)	(tons)	(tons)	(ton/mi²)
	Orthophosphate, in mg/L [00671]						
1	Big Sioux River at Akron, Iowa	973	421	43	397	2,010	0.139
2	Little Sioux River at Turin, Iowa	483	148	31	256	832	0.136
3	Boyer River at Logan, Iowa	155	10.7	6.9	135	177	0.178
4	Nishnabotna River above Hamburg, Iowa	329	15.1	4.6	301	360	0.117
5	Des Moines River at Keosauqua, Iowa	1,830	--	--	--	--	0.131
6	Skunk River at Augusta, Iowa	718	--	--	--	--	0.166
7	Iowa River at Wapello, Iowa	1,710	--	--	--	--	0.137
8	Wapsipinicon River near De Witt, Iowa	319	84.9	27	185	515	0.137
9	Maquoketa River near Spragueville, Iowa	179	30.0	17	127	245	0.110
10	Turkey River at Garber, Iowa	216	53.2	25	130	337	0.139
	Total	**6,910**	--	--	--	--	**0.137**
	Suspended sediment, in mg/L [80154]						
1	Big Sioux River at Akron, Iowa	608,000	84,500	14	459,000	790,000	86.9
2	Little Sioux River at Turin, Iowa	2,550,000	400,000	16	1,860,000	3,420,000	719
3	Boyer River at Logan, Iowa	1,890,000	792,000	42	791,000	3,830,000	2,170
4	Nishnabotna River above Hamburg, Iowa	3,320,000	631,000	19	2,260,000	4,720,000	1,180
5	Des Moines River at Keosauqua, Iowa	3,380,000	1,040,000	31	1,800,000	5,810,000	241
6	Skunk River at Augusta, Iowa	2,420,000	595,000	25	1,460,000	3,780,000	561
7	Iowa River at Wapello, Iowa	2,620,000	368,000	14	1,970,000	3,410,000	209
8	Wapsipinicon River near De Witt, Iowa	498,000	76,900	15	365,000	665,000	213
9	Maquoketa River near Spragueville, Iowa	437,000	68,400	16	318,000	586,000	267
10	Turkey River at Garber, Iowa	714,000	108,000	15	525,000	948,000	459
	Total	**18,400,000**	**4,160,000**	**23**	**11,800,000**	**28,000,000**	**364**

Table 6. Estimated major ions, nitrogen, phosphorus, and suspended-sediment loads and yields from selected major Iowa rivers, water years 2004–2008.—Continued

[Water year from October 1 to September 30; ID, identifier; SEP, standard error of prediction; L95 and U95, lower (L) and upper (U) limits of the 95-percent confidence interval of estimated load; ton/mi², tons per square mile; mg/L, milligrams per liter; parameter code given in brackets, see also table 2; μg/L, micrograms per liter; --, not available]

Map ID (fig. 2)	Station name	2008					
		Load	SEP		L95	U95	Yield
		(tons)	(tons)	(percent)	(tons)	(tons)	(ton/mi²)
	Chloride, in mg/L [00940]						
1	Big Sioux River at Akron, Iowa	63,100	1,780	2.8	59,700	66,600	9.02
2	Little Sioux River at Turin, Iowa	52,900	1,630	3.1	49,800	56,200	14.9
3	Boyer River at Logan, Iowa	12,700	389	3.1	12,000	13,500	14.7
4	Nishnabotna River above Hamburg, Iowa	36,900	1,070	2.9	34,900	39,100	13.1
5	Des Moines River at Keosauqua, Iowa	328,000	13,300	4.1	303,000	355,000	23.4
6	Skunk River at Augusta, Iowa	103,000	3,550	3.4	96,400	110,000	23.9
7	Iowa River at Wapello, Iowa	348,000	9,210	2.6	331,000	367,000	27.9
8	Wapsipinicon River near De Witt, Iowa	66,700	1,860	2.8	63,200	70,400	28.6
9	Maquoketa River near Spragueville, Iowa	43,400	1,330	3.1	40,800	46,100	26.6
10	Turkey River at Garber, Iowa	39,100	1,460	3.7	36,400	42,100	25.2
	Total	**1,090,000**	**35,500**	**3.3**	**1,030,000**	**1,170,000**	**21.6**
	Sulfate, in mg/L [00945]						
1	Big Sioux River at Akron, Iowa	401,000	18,300	4.6	366,000	438,000	57.3
2	Little Sioux River at Turin, Iowa	147,000	3,820	2.6	139,000	154,000	41.3
3	Boyer River at Logan, Iowa	24,800	833	3.4	23,200	26,400	28.5
4	Nishnabotna River above Hamburg, Iowa	73,100	2,300	3.1	68,700	77,700	26.0
5	Des Moines River at Keosauqua, Iowa	612,000	27,700	4.5	560,000	668,000	43.7
6	Skunk River at Augusta, Iowa	140,000	4,510	3.2	132,000	149,000	32.5
7	Iowa River at Wapello, Iowa	438,000	9,090	2.1	420,000	456,000	35.0
8	Wapsipinicon River near De Witt, Iowa	72,700	2,210	3.0	68,400	77,100	31.1
9	Maquoketa River near Spragueville, Iowa	51,000	1,260	2.5	48,600	53,500	31.2
10	Turkey River at Garber, Iowa	47,700	952	2.0	45,800	49,600	30.7
	Total	**2,010,000**	**71,000**	**3.5**	**1,870,000**	**2,150,000**	**39.8**
	Silica, in mg/L [00955]						
1	Big Sioux River at Akron, Iowa	36,000	--	--	--	--	5.15
2	Little Sioux River at Turin, Iowa	49,400	4,590	9.3	41,000	59,000	13.9
3	Boyer River at Logan, Iowa	12,500	735	5.9	11,200	14,000	14.4
4	Nishnabotna River above Hamburg, Iowa	54,800	--	--	--	--	19.5
5	Des Moines River at Keosauqua, Iowa	312,000	18,600	6.0	277,000	350,000	22.2
6	Skunk River at Augusta, Iowa	79,700	--	--	--	--	18.5
7	Iowa River at Wapello, Iowa	302,000	--	--	--	--	24.2
8	Wapsipinicon River near De Witt, Iowa	54,300	--	--	--	--	23.2
9	Maquoketa River near Spragueville, Iowa	34,600	3,080	8.9	29,000	41,100	21.2
10	Turkey River at Garber, Iowa	32,900	4,220	13	25,400	41,900	21.2
	Total	**968,000**	**--**	**--**	**--**	**--**	**19.1**

Table 6. Estimated major ions, nitrogen, phosphorus, and suspended-sediment loads and yields from selected major Iowa rivers, water years 2004–2008.—Continued

[Water year from October 1 to September 30; ID, identifier; SEP, standard error of prediction; L95 and U95, lower (L) and upper (U) limits of the 95-percent confidence interval of estimated load; ton/mi², tons per square mile; mg/L, milligrams per liter; parameter code given in brackets, see also table 2; µg/L, micrograms per liter; --, not available]

Map ID (fig. 2)	Station name	2008					
		Load	SEP		L95	U95	Yield
		(tons)	(tons)	(percent)	(tons)	(tons)	(ton/mi²)
Total nitrogen, in mg/L [49570 plus 62854, or 62855]							
1	Big Sioux River at Akron, Iowa	17,100	798	4.7	15,600	18,700	2.44
2	Little Sioux River at Turin, Iowa	29,400	1,380	4.7	26,800	32,200	8.29
3	Boyer River at Logan, Iowa	11,100	758	6.8	9,670	12,600	12.7
4	Nishnabotna River above Hamburg, Iowa	33,100	1,550	4.7	30,100	36,200	11.8
5	Des Moines River at Keosauqua, Iowa	114,000	7,820	6.9	99,200	130,000	8.11
6	Skunk River at Augusta, Iowa	50,000	5,100	10	40,700	60,700	11.6
7	Iowa River at Wapello, Iowa	145,000	51,800	36	69,100	269,000	11.6
8	Wapsipinicon River near De Witt, Iowa	43,000	5,590	13	33,100	54,900	18.4
9	Maquoketa River near Spragueville, Iowa	31,600	1,760	5.6	28,300	35,200	19.3
10	Turkey River at Garber, Iowa	29,300	1,760	6.0	26,000	32,900	18.9
	Total	**503,000**	**78,300**	**16**	**379,000**	**682,000**	**9.95**
Nitrate plus nitrite, in mg/L [00631]							
1	Big Sioux River at Akron, Iowa	14,000	1,190	8.5	11,800	16,500	2.00
2	Little Sioux River at Turin, Iowa	24,500	1,790	7.3	21,200	28,200	6.90
3	Boyer River at Logan, Iowa	7,090	483	6.8	6,190	8,080	8.15
4	Nishnabotna River above Hamburg, Iowa	24,500	1,750	7.1	21,200	28,100	8.71
5	Des Moines River at Keosauqua, Iowa	96,600	8,270	8.6	81,400	114,000	6.89
6	Skunk River at Augusta, Iowa	39,200	4,230	10.8	31,600	48,100	9.08
7	Iowa River at Wapello, Iowa	123,000	10,700	8.7	104,000	145,000	9.85
8	Wapsipinicon River near De Witt, Iowa	41,900	6,130	15	31,200	55,100	17.9
9	Maquoketa River near Spragueville, Iowa	24,300	1,660	6.8	21,200	27,800	14.9
10	Turkey River at Garber, Iowa	22,000	1,560	7.1	19,100	25,200	14.2
	Total	**417,000**	**37,700**	**9.0**	**348,000**	**496,000**	**8.25**
Total phosphorus, in mg/L [00665]							
1	Big Sioux River at Akron, Iowa	951	59.3	6.2	840	1,070	0.136
2	Little Sioux River at Turin, Iowa	1,530	231	15	1,130	2,030	0.431
3	Boyer River at Logan, Iowa	1,190	314	26	694	1,910	1.37
4	Nishnabotna River above Hamburg, Iowa	3,230	355	11	2,590	3,980	1.15
5	Des Moines River at Keosauqua, Iowa	5,880	329	5.6	5,260	6,550	0.419
6	Skunk River at Augusta, Iowa	3,800	484	13	2,940	4,830	0.880
7	Iowa River at Wapello, Iowa	7,190	330	4.6	6,570	7,860	0.575
8	Wapsipinicon River near De Witt, Iowa	1,230	187	15	909	1,640	0.528
9	Maquoketa River near Spragueville, Iowa	2,260	280	12	1,760	2,860	1.38
10	Turkey River at Garber, Iowa	1,570	430	27	894	2,560	1.01
	Total	**28,800**	**3,000**	**10**	**23,600**	**35,300**	**0.570**

Table 6. Estimated major ions, nitrogen, phosphorus, and suspended-sediment loads and yields from selected major Iowa rivers, water years 2004–2008.—Continued

[Water year from October 1 to September 30; ID, identifier; SEP, standard error of prediction; L95 and U95, lower (L) and upper (U) limits of the 95-percent confidence interval of estimated load; ton/mi², tons per square mile; mg/L, milligrams per liter; parameter code given in brackets, see also table 2; µg/L, micrograms per liter; --, not available]

Map ID (fig. 2)	Station name	2008					
		Load (tons)	SEP (tons)	(percent)	L95 (tons)	U95 (tons)	Yield (ton/mi²)
	Orthophosphate, in mg/L [00671]						
1	Big Sioux River at Akron, Iowa	326	106	33	167	577	0.0466
2	Little Sioux River at Turin, Iowa	517	120	23	322	788	0.146
3	Boyer River at Logan, Iowa	201	16.5	8.2	170	235	0.231
4	Nishnabotna River above Hamburg, Iowa	450	22.5	5.0	408	496	0.160
5	Des Moines River at Keosauqua, Iowa	3,030	--	--	--	--	0.216
6	Skunk River at Augusta, Iowa	1,380	--	--	--	--	0.319
7	Iowa River at Wapello, Iowa	3,300	--	--	--	--	0.264
8	Wapsipinicon River near De Witt, Iowa	859	332	39	386	1,660	0.368
9	Maquoketa River near Spragueville, Iowa	651	158	24	396	1,010	0.398
10	Turkey River at Garber, Iowa	447	128	29	247	746	0.287
	Total	**11,200**	--	--	--	--	**0.221**
	Suspended sediment, in mg/L [80154]						
1	Big Sioux River at Akron, Iowa	663,000	75,700	11	527,000	824,000	94.8
2	Little Sioux River at Turin, Iowa	2,150,000	323,000	15	1,590,000	2,850,000	605
3	Boyer River at Logan, Iowa	3,850,000	1,870,000	49	1,410,000	8,530,000	4,430
4	Nishnabotna River above Hamburg, Iowa	4,550,000	873,000	19	3,070,000	6,480,000	1,620
5	Des Moines River at Keosauqua, Iowa	7,310,000	2,420,000	33	3,700,000	13,100,000	521
6	Skunk River at Augusta, Iowa	4,170,000	1,320,000	32	2,160,000	7,280,000	965
7	Iowa River at Wapello, Iowa	3,960,000	637,000	16	2,860,000	5,340,000	317
8	Wapsipinicon River near De Witt, Iowa	754,000	145,000	19	511,000	1,080,000	323
9	Maquoketa River near Spragueville, Iowa	3,060,000	696,000	23	1,920,000	4,640,000	1,870
10	Turkey River at Garber, Iowa	2,840,000	643,000	23	1,790,000	4,290,000	1,830
	Total	**33,300,000**	**9,000,000**	**27**	**19,500,000**	**54,400,000**	**659**

Table 7. Regression models for estimating major ion, nutrient, and suspended-sediment loads from selected major Iowa rivers, water years 2004–2008.

[Water year from October 1 to September 30; ID, identifier; N, number; obs, observations; R^2, coefficient of determination; %, percent; mg/L, milligrams per liter; parameter code given in brackets, see also table 2; ln, natural logarithm; L, daily load in tons per day; Q, centered mean daily streamflow in cubic feet per second; T, centered time in decimal years; SS, seasonality parameter (2π*decimal years); AMLE, adjusted maximum likelihood estimation; dQ_i, change in flow relative to average flow of previous i number of days; A5yr, 5-year flow anomaly; A1yr, 1-year flow anomaly; A3mo, 3-month flow anomaly; HFV, high-frequency flow anomaly; LAD, least absolute deviation; BpQ, streamflow breakpoint term]

Map ID (fig. 2)	River	N. obs.	N. censored	Regression model	Estimated residual variance	R^2 (%)	Method[1]
				Chloride, in mg/L [00940]			
1	Big Sioux	56	0	$\ln(L) = 5.19 + 0.711{*}\ln Q - 0.0385{*}\ln Q^2 + 0.00820{*}T + 0.0161{*}T^2 + 0.115{*}\sin(SS) + 0.0970{*}\cos(SS)$	0.009	99	AMLE
2	Little Sioux	58	0	$\ln(L) = 4.77 + 0.865{*}\ln Q - 0.0505{*}\ln Q^2 + 0.0766{*}\sin(SS) + 0.108{*}\cos(SS) - 0.471{*}|dQ_1| - 0.101{*}dQ_{30}$	0.024	97	AMLE
3	Boyer	54	0	$\ln(L) = 3.62 + 0.676{*}\ln Q - 0.0186{*}\ln Q^2 - 0.253{*}dQ_1$	0.022	97	AMLE
4	Nishnabotna	58	0	$\ln(L) = 4.43 + 0.734{*}\ln Q - 0.0394{*}\ln Q^2 + 0.103{*}\sin(SS) + 0.0382{*}\cos(SS)$	0.023	98	AMLE
5	Des Moines	55	0	$\ln(L) = 6.13 + 0.117{*}\sin(SS) + 0.0524{*}\cos(SS) + (0.924{*}A5yr + 0.415{*}A1yr + 0.659{*}A3mo + 0.753{*}HFV)$	0.024	97	AMLE
6	Skunk	55	0	$\ln(L) = 4.87 + 0.735{*}\ln Q - 0.0335{*}\ln Q^2 + 0.146{*}\sin(SS) + 0.0957{*}\cos(SS)$	0.022	98	AMLE
7	Iowa	60	0	$\ln(L) = 6.31 + 0.0718{*}\sin(SS) + 0.0798{*}\cos(SS) + (0.404{*}A1yr + 0.711{*}A3mo + 0.699{*}HFV)$	0.011	98	AMLE
8	Wapsipinicon	55	0	$\ln(L) = 4.97 + 0.939{*}\ln Q - 0.0691{*}\ln Q^2 - 0.0409{*}T - 0.0571{*}T^2 + 0.0426{*}\sin(SS) + 0.0822{*}\cos(SS)$	0.009	99	AMLE
9	Maquoketa	56	0	$\ln(L) = 4.56 + 0.837{*}\ln Q - 0.135{*}\ln Q^2 + 0.0632{*}\sin(SS) + 0.0403{*}\cos(SS)$	0.016	98	AMLE
10	Turkey	55	0	$\ln(L) = 4.68 + 0.827{*}\ln Q - 0.0639{*}\ln Q^2$	0.026	97	AMLE
				Sulfate, in mg/L [00945]			
1	Big Sioux	57	0	$\ln(L) = 6.94 + 0.974{*}\ln Q + 0.0428{*}T - 0.227{*}dQ_{30}$	0.030	96	AMLE
2	Little Sioux	58	0	$\ln(L) = 5.79 + 0.791{*}\ln Q + 0.0666{*}\sin(SS) + 0.0890{*}\cos(SS) - 0.523{*}dQ_{30} - 0.195{*}|dQ_1|$	0.018	97	AMLE
3	Boyer	57	0	$\ln(L) = 4.26 + 0.739{*}\ln Q - 0.225{*}dQ_1$	0.020	98	AMLE
4	Nishnabotna	58	0	$\ln(L) = 5.04 + 0.784{*}\ln Q + 0.105{*}\sin(SS) + 0.0188{*}\cos(SS) - 0.165{*}dQ_{30}$	0.019	98	AMLE
5	Des Moines	56	0	$\ln(L) = 7.07 + 0.608{*}\ln Q - 0.0481{*}\ln Q^2 + 0.0964{*}\sin(SS) + 0.0217{*}\cos(SS) + 0.223{*}dQ_{30}$	0.044	94	AMLE
6	Skunk	56	0	$\ln(L) = 5.22 + 0.704{*}\ln Q - 0.0300{*}\ln Q^2 + 0.107{*}\sin(SS) + 0.0971{*}\cos(SS)$	0.020	98	AMLE
7	Iowa	60	0	$\ln(L) = 6.97 + 0.673{*}\ln Q - 0.0447{*}\ln Q^2 - 0.668{*}|dQ_1|$	0.012	97	AMLE
8	Wapsipinicon	57	0	$\ln(L) = 4.92 + 0.801{*}\ln Q - 0.0722{*}\ln Q^2$	0.018	98	AMLE
9	Maquoketa	56	0	$\ln(L) = 4.76 + 0.741{*}\ln Q - 0.112{*}\ln Q^2$	0.012	98	AMLE
10	Turkey	56	0	$\ln(L) = 4.56 + 0.837{*}\ln Q - 0.135{*}\ln Q^2 + 0.0632{*}\sin(SS) + 0.0403{*}\cos(SS)$	0.016	98	AMLE

Table 7. Regression models for estimating major ion, nutrient, and suspended-sediment loads from selected major Iowa rivers, water years 2004–2008.—Continued

[Water year from October 1 to September 30; ID, identifier; N., number; obs., observations; R^2, coefficient of determination; %, percent; mg/L, milligrams per liter; parameter code given in brackets, see also table 2; ln, natural logarithm; L, daily load in tons per day; Q, centered mean daily streamflow in cubic feet per second; T, centered time in decimal years; SS, seasonality parameter (2π*decimal years); AMLE, adjusted maximum likelihood estimation; dQ_i, change in flow relative to average flow of previous i number of days: A5yr, 5-year flow anomaly; A1yr, 1-year flow anomaly; A3mo, 3-month flow anomaly; HFV, high-frequency flow anomaly; LAD, least absolute deviation; BpQ, streamflow breakpoint term]

Map ID (fig. 2)	River	N. obs.	N. censored	Regression model	Estimated residual variance	R^2 (%)	Method[1]
				Silica, in mg/L [00955]			
1	Big Sioux	58	0	$\ln(L) = 4.49 + 1.21*\ln Q + 0.0852*T - 0.126*\sin(SS) + 0.159*\cos(SS)$	0.258	86	LAD
2	Little Sioux	58	0	$\ln(L) = 4.64 + 1.24*\ln Q - 0.225*\ln Q^2 + 0.0888*T - 0.0699*T^2 - 0.125*\sin(SS) + 0.189*\cos(SS)$	0.105	94	AMLE
3	Boyer	56	0	$\ln(L) = 3.26 + 1.03*\ln Q - 0.0754*\ln Q^2 - 0.338*dQ_1$	0.065	96	AMLE
4	Nishnabotna	57	0	$\ln(L) = 4.51 + 1.07*\ln Q - 0.0401*\ln Q^2 + 0.0309*T - 0.196*dQ_{30}$	0.022	96	LAD
5	Des Moines	53	0	$\ln(L) = 5.63 + 1.26*\ln Q - 0.132*\sin(SS) + 0.209*\cos(SS) - 0.221*dQ_{30}$	0.022	97	AMLE
6	Skunk	51	0	$\ln(L) = 4.12 + 1.36*\ln Q - 0.205*\ln Q^2 - 0.271*\sin(SS) + 0.0256*\cos(SS)$	0.376	88	LAD
7	Iowa	28	0	$\ln(L) = 5.89 + 1.24*\ln Q$	0.241	83	LAD
8	Wapsipinicon	54	0	$\ln(L) = 3.58 + 1.52*\ln Q - 0.239*\sin(SS) + 0.585*\cos(SS)$	1.018	71	LAD
9	Maquoketa	56	0	$\ln(L) = 3.68 + 0.923*\ln Q + 0.0901*T + 0.0734*T^2 - 0.0511*\sin(SS) + 0.142*\cos(SS)$	0.082	92	AMLE
10	Turkey	55	0	$\ln(L) = 4.15 + 1.15*\ln Q - 0.105*\ln Q^2 + 0.0846*T$	0.165	91	AMLE
				Total nitrogen, in mg/L [49570 plus 62854, or 62855]			
1	Big Sioux	58	0	$\ln(L) = 3.92 + 1.09*\ln Q - 0.0817*\ln Q^2 + 0.174*\sin(SS) + 0.172*\cos(SS)$	0.064	96	AMLE
2	Little Sioux	58	0	$\ln(L) = 3.87 + 1.31*\ln Q - 0.111*\ln Q^2 + 0.221*\sin(SS) + 0.114*\cos(SS)$	0.057	97	AMLE
3	Boyer	57	0	$\ln(L) = 3.17 + 1.12*\ln Q - 0.0513*\ln Q^2 + 0.128*\sin(SS) - 0.0142*\cos(SS)$	0.078	97	AMLE
4	Nishnabotna	58	0	$\ln(L) = 4.09 + 1.22*\ln Q - 0.159*\ln Q^2 + 0.158*\sin(SS) - 0.0403*\cos(SS)$	0.047	98	AMLE
5	Des Moines	55	0	$\ln(L) = 5.25 + 1.35*\ln Q - 0.124*T - 0.132*T^2 + 0.315*\sin(SS) + 0.0704*\cos(SS) - 0.345*dQ_{30}$	0.061	97	AMLE
6	Skunk	56	0	$\ln(L) = 3.70 + 1.22*\ln Q - 0.181*\ln Q^2 + 0.359*\sin(SS) - 0.0153*\cos(SS)$	0.167	95	AMLE
7	Iowa	59	0	$\ln(L) = 5.99 + 1.33*\ln Q - 0.142*\ln Q^2 - 0.121*T - 0.0841*T^2 + 0.114*\sin(SS) + 0.121*\cos(SS) - 0.556*dQ_1$	0.041	97	AMLE
8	Wapsipinicon	55	0	$\ln(L) = 3.66 + 1.41*\ln Q - 0.151*\ln Q^2 + 0.147*\sin(SS) + 0.224*\cos(SS)$	0.101	95	AMLE
9	Maquoketa	56	0	$\ln(L) = 3.89 + 1.20*\ln Q - 0.0935*\ln Q^2 + 0.0904*\sin(SS) + 0.0902*\cos(SS)$	0.038	97	AMLE
10	Turkey	55	0	$\ln(L) = 4.09 + 1.21*\ln Q - 0.0528*\ln Q^2 + 0.0697*\sin(SS) + 0.0798*\cos(SS)$	0.042	98	AMLE

Table 7. Regression models for estimating major ion, nutrient, and suspended-sediment loads from selected major Iowa rivers, water years 2004–2008.—Continued

[Water year from October 1 to September 30; ID, identifier; N., number, obs., observations; R^2, coefficient of determination; %, percent; mg/L, milligrams per liter, parameter code given in brackets, see also table 2; ln, natural logarithm; L, daily load in tons per day; Q, centered mean daily streamflow in cubic feet per second; T, centered time in decimal years; SS, seasonality parameter (2π*decimal years); AMLE, adjusted maximum likelihood estimation; dQ_i, change in flow relative to average flow of previous i number of days; A5yr, 5-year flow anomaly; A1yr, 1-year flow anomaly; A3mo, 3-month flow anomaly; HFV, high-frequency flow anomaly; LAD, least absolute deviation; BpQ, streamflow breakpoint term]

Map ID (fig. 2)	River	N. obs.	N. censored	Regression model	Estimated residual variance	R^2 (%)	Method[1]
				Nitrate plus nitrite, in mg/L [00631]			
1	Big Sioux	58	0	$\ln(L) = 3.72 + 1.24*\ln Q - 0.122*\ln Q^2 + 0.300*\sin(SS) + 0.439*\cos(SS) - 0.301*dQ_{30}$	0.191	89	AMLE
2	Little Sioux	57	0	$\ln(L) = 3.73 + 1.45*\ln Q - 0.179*\ln Q^2 + 0.269*\sin(SS) + 0.362*\cos(SS) - 0.390*dQ_{30}$	0.122	94	AMLE
3	Boyer	57	0	$\ln(L) = 2.92 + 0.985*\ln Q - 0.0731*\ln Q^2 - 0.398*dQ_1d$	0.087	95	AMLE
4	Nishnabotna	58	0	$\ln(L) = 3.88 + 1.32*\ln Q - 0.125*\ln Q^2 - 0.0632*T + 0.166*\sin(SS) + 0.0867*\cos(SS) - 0.397*dQ_{30}$	0.068	97	AMLE
5	Des Moines	56	0	$\ln(L) = 5.12 + 1.56*\ln Q - 0.171*T - 0.160*T^2 + 0.286*\sin(SS) + 0.236*\cos(SS) - 0.657*dQ_{30}$	0.094	96	AMLE
6	Skunk	53	6	$\ln(L) = 8.47 + 6.37*\ln Q - 5.63*BpQ$	0.205	98	AMLE
7	Iowa	57	0	$\ln(L) = 6.08 + 1.45*\ln Q - 0.147*\ln Q^2 - 0.153*T - 0.101*T^2 + 0.0925*\sin(SS) + 0.264*\cos(SS) - 1.42*|dQ_1|$	0.089	94	AMLE
8	Wapsipinicon	55	0	$\ln(L) = 3.81 + 1.52*\ln Q - 0.212*\ln Q^2 + 0.210*\sin(SS) + 0.437*\cos(SS)$	0.245	92	AMLE
9	Maquoketa	55	0	$\ln(L) = 3.73 + 1.20*\ln Q - 0.104*\ln Q^2 + 0.0757*\sin(SS) + 0.189*\cos(SS) - 0.293*dQ_{30}$	0.054	96	AMLE
10	Turkey	54	0	$\ln(L) = 4.10 + 1.21*\ln Q - 0.0628*\ln Q^2 - 1.15*|dQ_1|$	0.065	96	AMLE
				Total phosphorus, in mg/L [00665]			
1	Big Sioux	58	0	$\ln(L) = 0.787 + 0.831*\ln Q + 0.0970*\sin(SS) + 0.215*\cos(SS) + 0.596*dQ_{30}$	0.102	94	AMLE
2	Little Sioux	57	0	$\ln(L) = 0.542 + 1.55*\ln Q - 0.108*T + 0.459*dQ_{30}$	0.251	92	AMLE
3	Boyer	57	0	$\ln(L) = 1.066 + 1.47*\ln Q + 0.0700*\ln Q^2 - 0.270*T - 0.177*T^2$	0.218	94	AMLE
4	Nishnabotna	57	0	$\ln(L) = 1.41 + 1.31*\ln Q - 0.140+\ln Q^2 - 0.115*T - 0.00113*\sin(SS) - 0.385*\cos(SS) + 0.638*dQ_{30}$	0.117	97	AMLE
5	Des Moines	54	0	$\ln(L) = 1.88 + 1.10*\ln Q + 0.123*\sin(SS) + 0.208*\cos(SS) + 0.221*dQ_{30}$	0.057	97	AMLE
6	Skunk	54	0	$\ln(L) = 0.659 + 1.17*\ln Q - 0.0685*\ln Q^2 + 0.00225*\sin(SS) - 0.344*\cos(SS) + 0.395*dQ_{30}$	0.144	96	AMLE
7	Iowa	59	0	$\ln(L) = 2.41 + 0.914*\ln Q + 0.0426*T + 0.0214*\sin(SS) - 0.174*\cos(SS) + 0.244*dQ_{30}$	0.030	98	AMLE
8	Wapsipinicon	56	0	$\ln(L) = 0.331 + 0.999*\ln Q - 0.175*\ln Q^2 - 0.0940*\sin(SS) - 0.626*\cos(SS) + 0.551*dQ_{30}$	0.146	93	AMLE
9	Maquoketa	55	0	$\ln(L) = 0.377 + 1.41*\ln Q + 0.0609*\sin(SS) - 0.189*\cos(SS) - 0.293*dQ_{30}$	0.098	96	AMLE
10	Turkey	55	0	$\ln(L) = 0.651 + 1.85*\ln Q - 0.0948*\ln Q^2 - 0.163*T$	0.339	92	AMLE

Table 7. Regression models for estimating major ion, nutrient, and suspended-sediment loads from selected major Iowa rivers, water years 2004–2008.—Continued

[Water year from October 1 to September 30; ID, identifier; N., number; obs., observations; R², coefficient of determination; %, percent; mg/L, milligrams per liter; parameter code given in brackets, see also table 2; ln, natural logarithm; L, daily load in tons per day; Q, centered mean daily streamflow in cubic feet per second; T, centered time in decimal years; SS, seasonality parameter (2π*decimal years); AMLE, adjusted maximum likelihood estimation; dQ$_i$, change in flow relative to average flow of previous i number of days; A5yr, 5-year flow anomaly; A1yr, 1-year flow anomaly; A3mo, 3-month flow anomaly; HFV, high-frequency flow anomaly; LAD, least absolute deviation; BpQ, streamflow breakpoint term]

Map ID (fig. 2)	River	N. obs.	N. censored	Regression model	Estimated residual variance	R² (%)	Method[1]
				Orthophosphate, in mg/L [00671]			
1	Big Sioux	57	0	$\ln(L) = -2.02 + 0.209*\sin(SS) + 0.935*\cos(SS) + (-0.574*A5yr + 1.77*A1yr + 1.20*A3mo + 1.69*HFV)$	0.940	79	AMLE
2	Little Sioux	58	7	$\ln(L) = -0.759 + 2.03*\ln Q - 0.437*\ln Q^2 + 0.460*\sin(SS) + 0.695*\cos(SS)$	0.929	84	AMLE
3	Boyer	57	0	$\ln(L) = -1.01 - 0.228*\sin(SS) - 0.0172*\cos(SS) + (1.22*A5yr + 0.358*A1yr + 0.485*A3mo + 0.732*HFV)$	0.085	90	AMLE
4	Nishnabotna	58	0	$\ln(L) = -0.143 + 0.938*\ln Q - 0.0860*\ln Q^2 - 0.0485*\sin(SS) - 0.226*\cos(SS)$	0.057	97	AMLE
5	Des Moines	55	0	$\ln(L) = 1.25 + 1.37*\ln Q + 0.0333*\sin(SS) + 0.810*\cos(SS) - 0.399*dQ_{30}$	0.416	81	LAD
6	Skunk	54	0	$\ln(L) = -0.486 + 1.38*\ln Q - 0.268*\sin(SS) + 0.146*\cos(SS)$	0.698	83	LAD
7	Iowa	57	0	$\ln(L) = 1.63 + 1.40*\ln Q + 0.00372*\sin(SS) + 0.746*\cos(SS)$	0.503	81	LAD
8	Wapsipinicon	57	12	$\ln(L) = -1.38 + 2.41*\ln Q - 0.158*\ln Q^2 - 0.250*\sin(SS) + 1.066*\cos(SS)$	0.798	87	AMLE
9	Maquoketa	56	2	$\ln(L) = -0.325 + 1.68*\ln Q - 0.203*\ln Q^2 - 0.181*\sin(SS) + 0.343*\cos(SS)$	0.531	81	AMLE
10	Turkey	55	6	$\ln(L) = -0.531 + 1.75*\ln Q - 0.197*\ln Q^2 + 0.127*\sin(SS) + 0.334*\cos(SS)$	0.644	85	AMLE
				Suspended sediment, in mg/L [80154]			
1	Big Sioux	57	0	$\ln(L) = 7.16 + 1.18*\ln Q - 0.134*\ln Q^2 - 0.0255*\sin(SS) - 0.417*\cos(SS) + 0.521*dQ_{30}$	0.259	91	AMLE
2	Little Sioux	56	0	$\ln(L) = 7.50 + 1.90*\ln Q - 0.110*T - 0.0938*\sin(SS) - 0.250*\cos(SS) + 0.383*dQ_{30}$	0.216	96	AMLE
3	Boyer	55	0	$\ln(L) = 7.43 + 2.12*\ln Q - 0.160*\ln Q^2 + 0.734*dQ_1$	0.535	95	AMLE
4	Nishnabotna	57	0	$\ln(L) = 8.41 + 1.65*\ln Q - 0.316*\ln Q^2 - 0.157*T - 0.117*\sin(SS) - 0.548*\cos(SS) + 0.907*dQ_{30}$	0.310	96	AMLE
5	Des Moines	55	0	$\ln(L) = 7.38 + 1.50*\ln Q + 0.0732*\sin(SS) - 0.534*\cos(SS) + 0.904*dQ_{30}$	0.810	87	AMLE
6	Skunk	55	0	$\ln(L) = 6.91 + 1.50*\ln Q - 0.144*\ln Q^2 - 0.0643*\sin(SS) - 0.544*\cos(SS) + 0.662*dQ_{30}$	0.548	91	AMLE
7	Iowa	57	0	$\ln(L) = 8.60 + 1.11*\ln Q - 0.276*\ln Q^2 - 0.0736*\sin(SS) - 0.561*\cos(SS) + 1.61*dQ_1 + 0.625*dQ_{30}$	0.331	88	AMLE
8	Wapsipinicon	53	0	$\ln(L) = 6.82 + 1.35*\ln Q - 0.258*\ln Q^2 - 0.298*\sin(SS) - 0.363*\cos(SS) + 3.03*dQ_1$	0.390	87	AMLE
9	Maquoketa	54	0	$\ln(L) = 5.93 - 0.0611*\sin(SS) - 0.568*\cos(SS) + (2.21*A1yr + 0.946*A3mo + 2.16*HFV)$	0.279	93	AMLE
10	Turkey	53	0	$\ln(L) = 4.90 - 0.284*\sin(SS) - 0.269*\cos(SS) + (2.53*A1yr + 1.67*A3mo + 1.95*HFV)$	0.192	95	AMLE

[1]LAD method does not produce estimated residual variance or R²; values are based on similar models using AMLE.

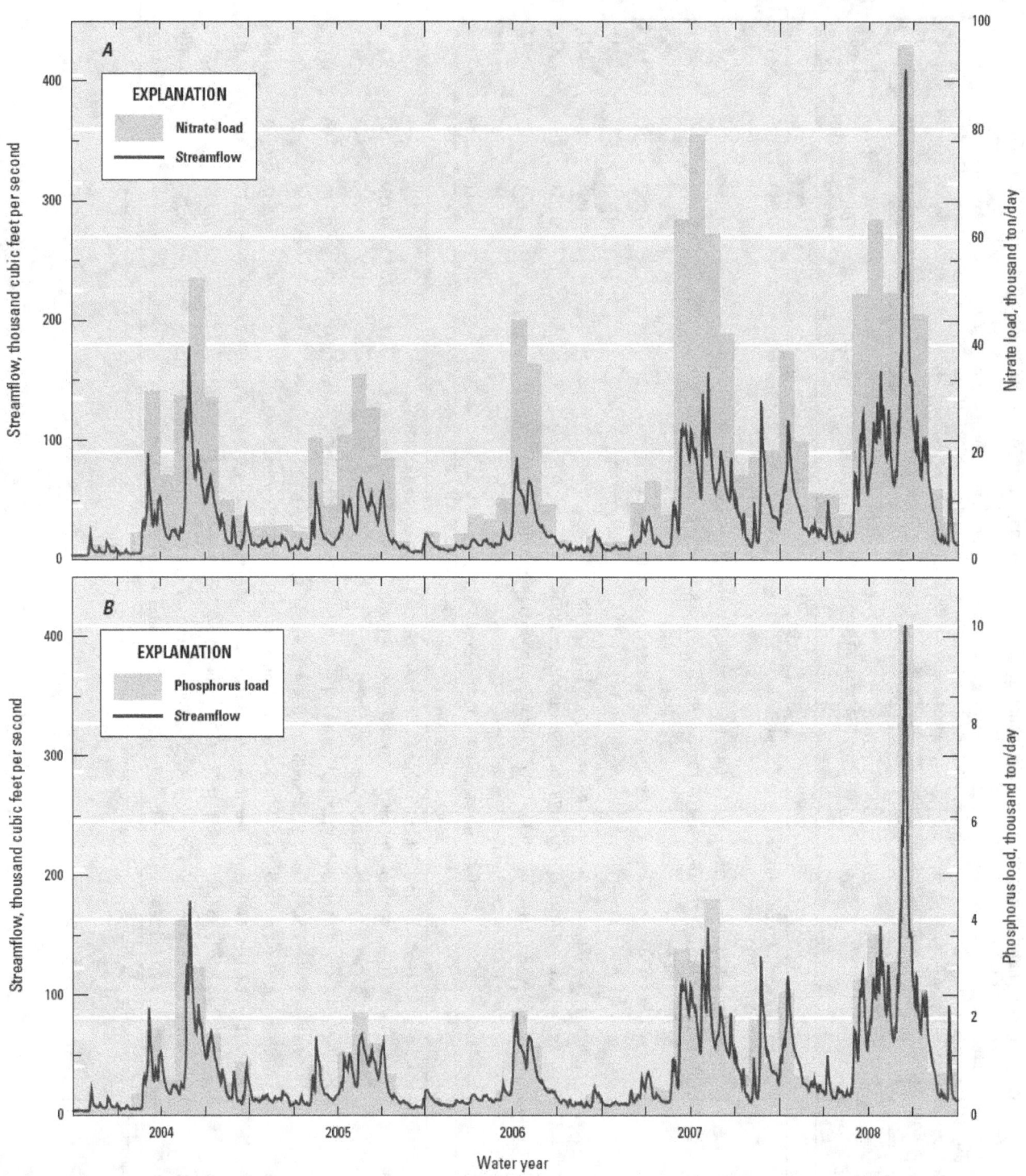

Figure 7.   Total nitrate and total phosphorus load and combined streamflow for 10 major Iowa rivers, water years 2004–2008.

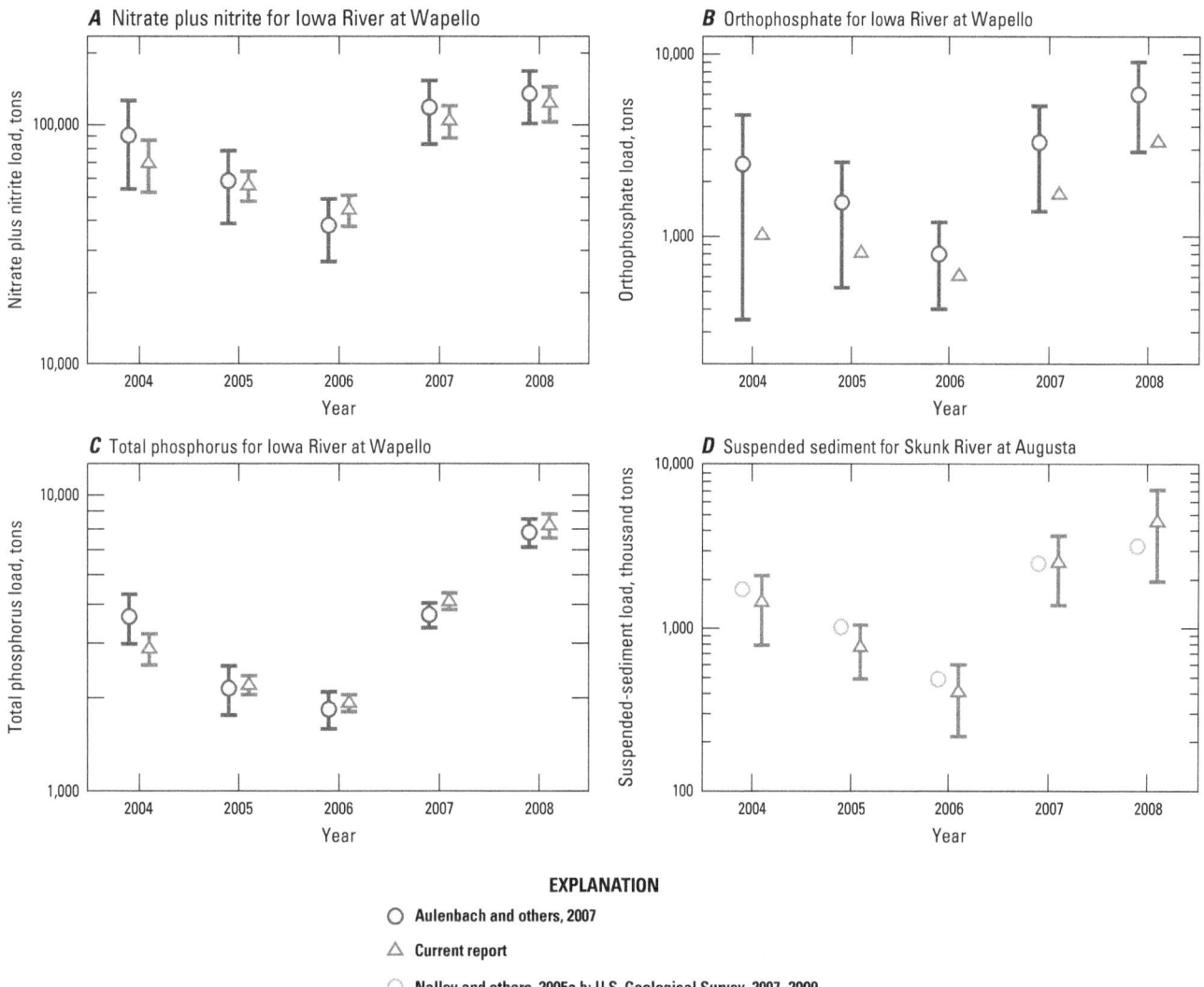

Figure 8. Comparison of annual load estimates and 95-percent confidence limits for select constituents at selected sites, water years 2004–2008.

Because TN concentrations in Iowa streams are dominated by nitrate, patterns in nitrate loads and yields are similar to TN. The smallest nitrate load was estimated at 1,040 tons in 2006 from the Boyer River (map ID 3), and the largest load was 123,000 tons in 2008 from the Iowa River (map ID 7; table 6). The Des Moines (map ID 5) and Iowa Rivers consistently had the largest nitrate loads of the studied rivers, and the Boyer River the smallest. Nitrate yields ranged from 0.723 ton/mi² in 2006 from the Nishnabotna River (map ID 4) to 17.9 ton/mi² in 2008 from the Iowa River. Similar to total nitrogen, average nitrate yields for the 5-year study period were smallest at the Big Sioux River and largest in northeastern basins (map IDs 7–10; fig. 9E). Total nitrogen and nitrate yields relative to a long-term average streamflow indicated that Mississippi tributaries yielded only slightly more nitrogen than Missouri tributaries, with the greatest yields related less

to site differences than to exceptionally wet conditions (years with greater than twice the long-term average streamflow).

The patterns in nitrogen loads are largely reflective of the relation with streamflow, but also demonstrate the seasonal effect of agricultural practices on the landscape. Nitrogen fertilizer application is greatest in the spring, coincident with planting and typical spring increases in precipitation and streamflow. Fertilizer distribution data, available biannually at the statewide level for the State of Iowa (Iowa Department of Agriculture and Land Stewardship, Fertilizer Tonnage Distribution in Iowa, *http://www.iowaagriculture.gov/ feedAndFertilizer/fertilizerDistributionReport.asp*, accessed February 18, 2009), compared with combined basin-wide loads for the 10 study basins (fig. 10), provides a useful, albeit incomplete, comparison of nutrient application and stream loads. Figure 10A shows that although nitrogen fertilizer

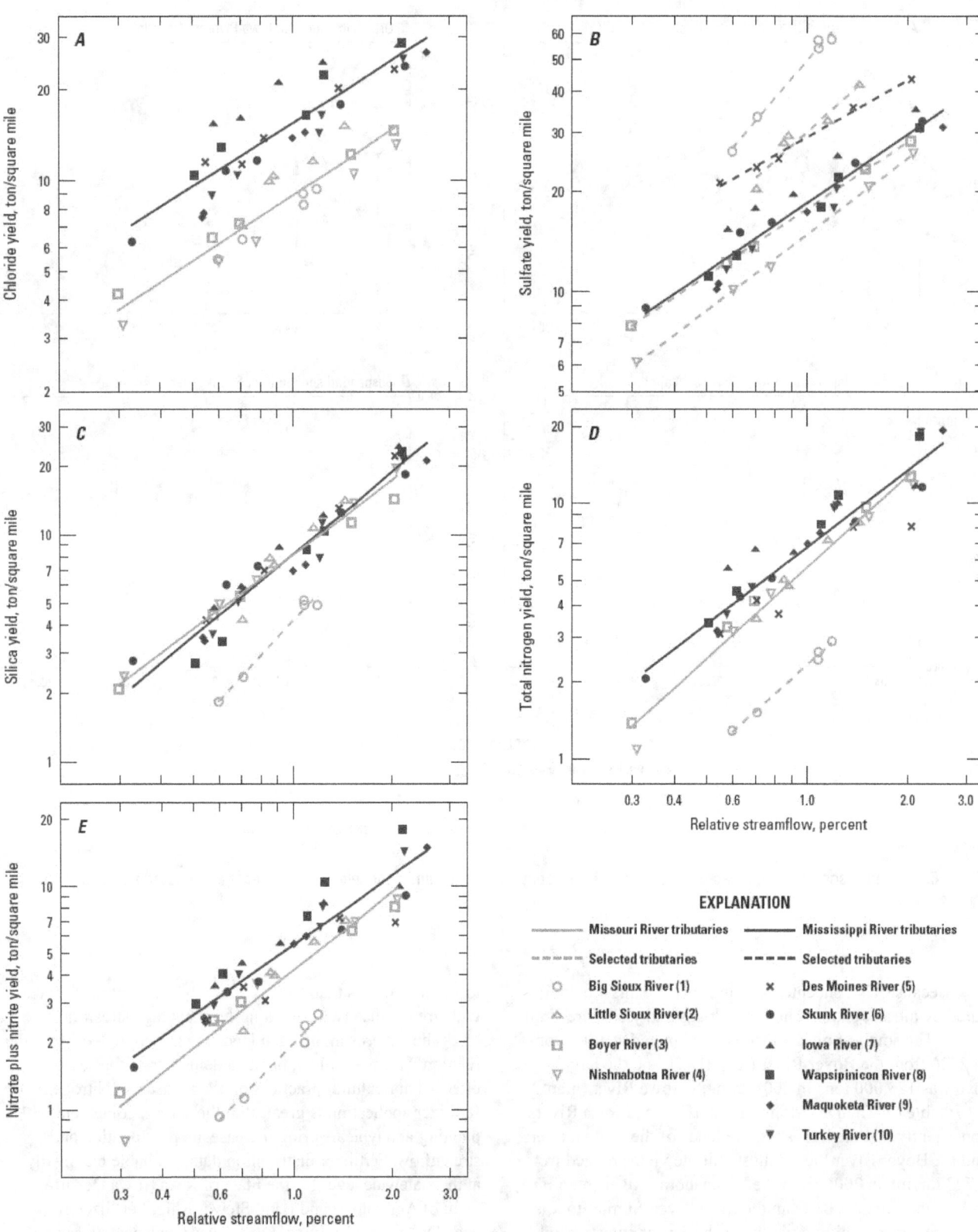

Figure 9. Relations (log-fit) between annual constituent yield and average streamflow relative to long-term (1979–2008) average streamflow for 10 major Iowa rivers showing map identifier, water years 2004–2008.

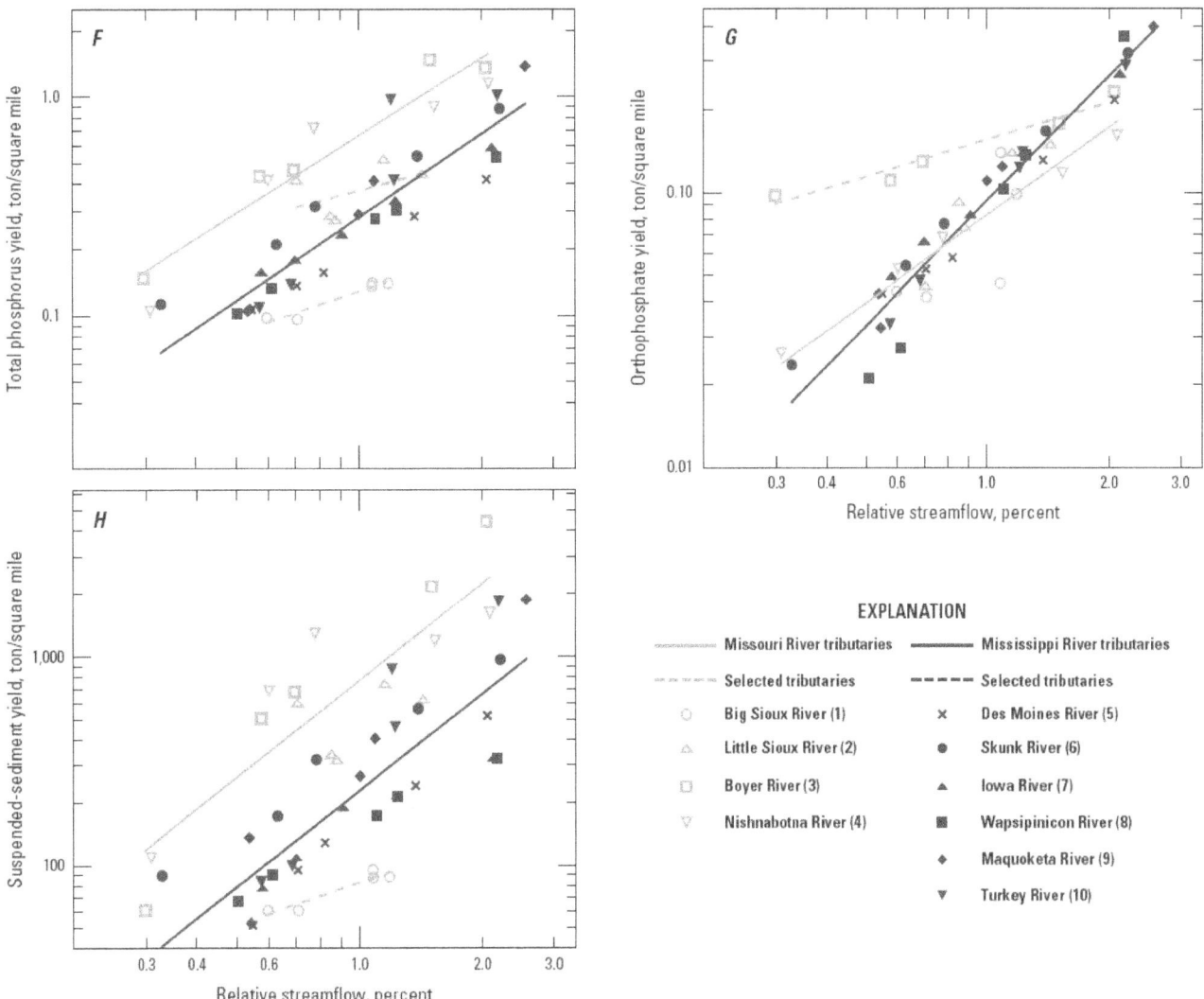

Figure 9. Relations (log-fit) between annual constituent yield and average streamflow relative to long-term (1979–2008) average streamflow for 10 major Iowa rivers showing map identifier, water years 2004–2008.—Continued

application (expressed as statewide distribution) and nitrogen loads are greater in the first one-half of the year, loads relative to fertilizer application are greater during wet periods. This pattern was evident in the dry years of 2004 through 2006, but was more dramatic in 2007 and 2008, when nitrogen stream loads were above 40 percent relative to the amount of nitrogen fertilizer application and above-average precipitation and streamflow was recorded. Nitrogen loads relative to fertilizer use for January to June (including the critical period of spring application and high streamflow) rise from about 30 percent in 2004–2006 to an average 52 percent in the wet 2007 and 2008 years. Loads relative to use were much smaller for nitrogen in the second one-half of the year, with July-to-December loads ranging from 8 to 27 percent.

Nutrient loads are presented relative to fertilizer distribution to provide perspective. Only 75 percent of the State of Iowa is included in the study basins and areas of Minnesota and South Dakota account for 17 percent of the study basins. Agricultural fertilizer certainly is not the only source of nutrients in streams, but the mass of nitrogen exported from the State of Iowa from the 10 study basins follows similar temporal patterns to distribution, and this load represents a large part of the statewide nitrogen use.

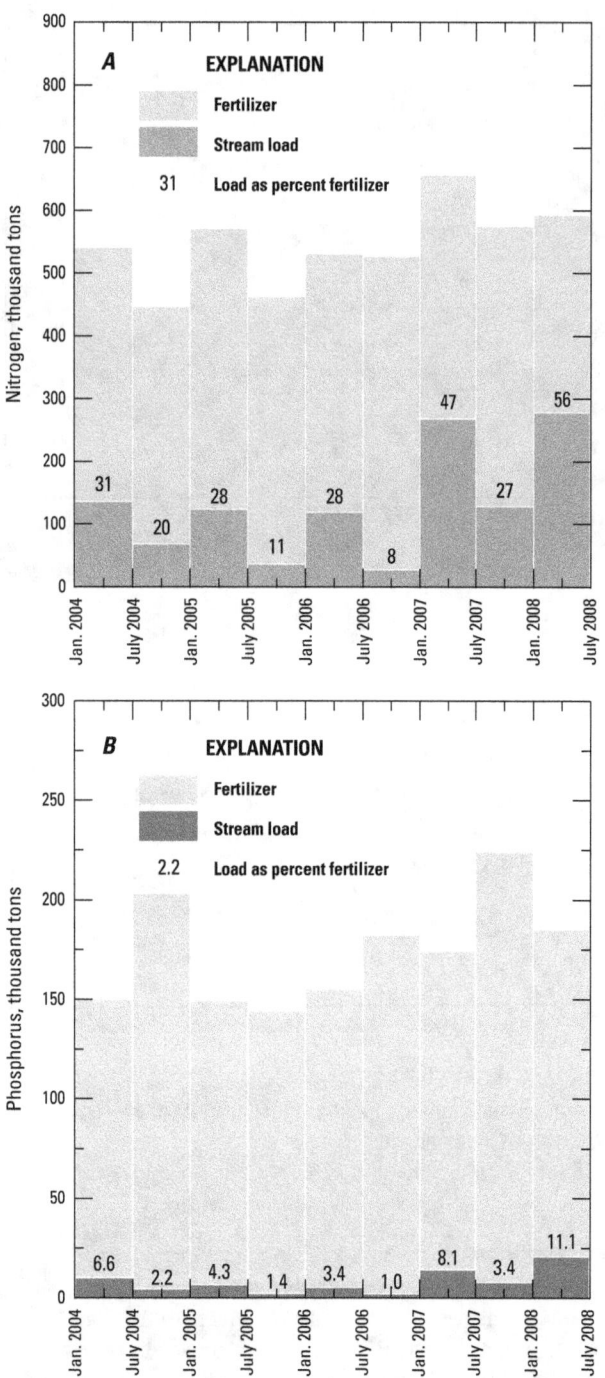

Figure 10. Fertilizer distribution in Iowa and constituent loads combined for 10 major Iowa rivers, 2004–2008.

Phosphorus

Total phosphorus loads ranged from 129 tons in 2006 from the Boyer River (map ID 3) to 7,190 tons in 2008 from the Iowa River (map ID 7; table 6). The effect of streamflow on loads resulted in larger TP loads from large central Iowa basins, particularly the Iowa and Des Moines (map ID 5) Rivers, than from smaller basins. The smallest annual TP yield was in 2005 from the Big Sioux River (map ID 1) at 0.0959 ton/mi², and the largest annual TP yield was in 2007 at 1.48 ton/mi² from the Boyer River (map ID 3). The Boyer and Nishnabotna Rivers (map IDs 3 and 4) reported results of consistently high TP yields each year, with an exception for the Nishnabotna River in 2006, which was a dry year across much of the State, as indicated by relatively lower streamflow compared to the long-term average for 1979–2008. Excepting the Nishnabotna River in 2006, the Big Sioux River had the lowest annual TP yields through the 5-year study period (fig. 9F).

Annual orthophosphate loads ranged from 49.0 tons in 2005 from the Wapsipinicon River (map ID 8) to 3,300 tons in 2008 from the Iowa River (map ID 7; table 6). The Iowa and Des Moines Rivers (map ID 5) had the greatest annual orthophosphate loads, generally twice the annual load of any other site, because of greater streamflow. The Big Sioux River also had high orthophosphate loads in 2006 and 2007. The annual orthophosphate yields among all 10 sites ranged from 0.0210 ton/mi² in 2005 from the Wapsipinicon River to 0.398 ton/mi² in 2008 from the Maquoketa River. The Boyer River was consistently one of the greatest orthophosphate-yielding basins relative to the other sites, particularly during dry years (fig. 9G).

The seasonal patterns in phosphorus loads, similar to nitrogen, are largely reflective of the relation with streamflow, but the effects of agricultural practices differ for nitrogen and phosphorus. Phosphorus commercial fertilizer applications are typically greatest in the fall after harvest when stream-flows are generally low. Fertilizer application expressed as statewide distribution, along with combined stream total phosphorus loads with an indication of percent loads relative to application is shown in figure 10B. As with nitrogen, loads are greater during wet periods than dry, with about a fourfold increase in loads between the driest (2006) and wettest water years (2008). Phosphorus January-to-June relative loads for 2004–2006 average 4.8 percent, and 9.6 percent for 2007–2008. Loads relative to use were much smaller for phosphorus in the second one-half of the year, with July-to-December loads ranging from 1.0 to 3.4 percent.

Suspended Sediment

Annual suspended-sediment loads ranged from 53,000 tons from the Boyer River (map ID 3) in 2006 to 7,310,000 tons from the Des Moines River (map ID 5) in 2008 (table 6). The general relation between basin size and constituent load does not hold well for suspended sediment.

The Nishnabotna River (map ID 4) loads are similar to the Des Moines River, although the Des Moines River basin is five times larger. Conversely, the Big Sioux (map ID 1) and Wapsipinicon Rivers (map ID 8), ranked three and seven by basin size, had consistently lower loads and yields of suspended sediment than other rivers. Annual yields ranged from 51.8 ton/mi² from the Des Moines River in 2006 to 4,430 ton/mi² from the Boyer River in 2008. The Big Sioux River generally had the lowest suspended-sediment yields, whereas other Missouri River tributaries, particularly the two southwestern Iowa basins of the Boyer and Nishnabotna Rivers tended to have greater suspended-sediment yields than Mississippi River tributaries (fig. 9H).

Summary

Concentrations, loads, and yields of streamflow constituents were assessed for 10 large Iowa rivers for the 2004 through 2008 water years including analysis of major ions, carbon, nutrients, suspended sediment, and pesticides. Fertilizers and pesticides are commonly used on agricultural, residential, and urban lands to sustain crop yields and support the desired urban landscape with lawns. Fertilizer and pesticide application and variable climate conditions can lead to precipitation washing some of these chemicals into streams and rivers. Pesticides and high levels of nutrients can have deleterious effects on aquatic health of streams, and nutrients transported from Iowa and other Midwest agricultural states have been linked to hypoxia in the Gulf of Mexico.

Samples were collected from March 2004 through September 2008 as part of a project in cooperation with the Iowa Department of Natural Resources. Sampling sites are located near the mouths of Iowa rivers in basins, which cover large parts of the State, ranging in size from 871 to 14,038 square miles, covering a total 50,562 square miles. Basins include 75.0 percent of Iowa and additional areas in eastern South Dakota and southern Minnesota accounting for 17.1 percent of the study basins. The average annual precipitation gradient increases across the basins from 22 to 38 inches from northwest to southeast. The variations in basin size and precipitation resulted in long-term (1979–2008) average annual streamflows at sampling sites from 461 to 10,293 cubic feet per second.

Iowa land use is largely agricultural, with 73 percent of the State used for crop production, 86 percent of which is corn and soybeans. Iowa is one of the most productive areas for corn and soybeans in the world. The glacial and alluvial landforms vary across the study basins, though most of this region of fertile, moist, glacial, commonly calcareous soils, and former prairies is well suited to extensive agricultural land uses. The thick loess deposits and steep to rolling hills of western and southern Iowa produce highly erodible slopes. Population density in this rural state averages 54 people per square mile.

Water-quality samples were collected using standard protocols to obtain streamflow-integrated samples, typically using equal-width increment techniques. The exclusion of methanol from cleaning procedures reduced occurrence of dissolved organic carbon from field blank quality-control samples without increased incident of pesticide carry-over. Samples sent to USGS laboratories for analysis of major ions, nutrients, carbon, pesticides, and suspended sediment.

Statistical summaries of sample data computed in TIBCO Spotfire S+® used nonparametric regression on order statistics, parametric adjusted maximum likelihood estimation methods, and a modification of the Kaplan-Meier nonparametric method. These methods provide correct handling of datasets with values below analytical detection limits and with changing levels of detection.

Stream loads, the chemical mass transported by a stream past a location during a specified time period, were estimated for water years 2004 through 2008 by a rating-curve method using S-LOADEST. Stream yields (loads divided by watershed area) were computed to compare constituent contributions from watersheds of different sizes. In addition to predefined models using linear and quadratic streamflow and time terms with sine and cosine to describe seasonal patterns, additional terms describing streamflow variability and anomalies were evaluated. Streamflow variability terms describe the difference in streamflow from recent average conditions, on a 1-day or 30-day time step. Streamflow anomaly terms account for deviations from average conditions sequentially from long- to short-term, using 5-year, 1-year, 3-month, and high frequency variation terms. Candidate regression load models were evaluated for model fit, distribution assumptions on the residuals, and correlation of the explanatory variables, with preferred models with low residual variance, normal and homoskedastic residual distributions, low correlation among explanatory variables, and good empirical agreement with measured data.

Constituent concentrations vary by streamflow and season in Iowa. Constituent concentrations decreased with streamflow for pH, alkalinity, specific conductance, chloride, and sulfate, whereas concentrations increased with streamflow for particulate and dissolved organic carbon, total phosphorus, and suspended sediment. Silica, particulate inorganic carbon, and chlorophyll-a concentrations did not correlate directly with streamflow. Nitrogen concentrations (total nitrogen and nitrate plus nitrite) increased with low and moderate streamflows, but decreased with high streamflows. The seasonality of streamflow affected concentrations, but additional seasonal patterns were defined by algae blooms and pesticide application.

Climate and landscape gradients resulted in spatial patterns across Iowa in specific conductance, alkalinity, chloride, total phosphorus, dissolved organic carbon, suspended sediment, and turbidity. Alkalinity was greatest in northern Iowa rivers. Chloride and dissolved organic carbon increased among tributaries upstream along the Missouri River and downstream among tributaries along the Mississippi River. Turbidity, suspended sediment, and total phosphorus were greatest from southwestern Iowa rivers draining loess landscapes. Specific conductance and sulfate concentrations were greatest and most variable in the Big Sioux River in northwestern Iowa. Spatial variability also was evident in different streamflow-concentration relations among sites for particulate organic carbon, orthophosphate, suspended sediment and turbidity. Spatial patterns were not evident for pH, silica, nitrogen (all forms), or algal pigments.

Major ion and carbon concentrations were largely reflective of the calcareous, glacial soils. In general, the rivers sampled in Iowa were alkaline and well-buffered, with pH and alkalinity inversely related to streamflow with the greatest alkalinities during long periods of stable streamflow. Alkalinities were greatest in northern Iowa streams, though no spatial pattern was observed for pH levels. Specific conductance and ion concentrations were also inversely related to streamflow, except silica, which was not related to streamflow. Specific conductance and sulfate concentrations were greatest in the Big Sioux River. Carbon in streamwater was dominated by bicarbonate, followed by particulate organic carbon and dissolved organic carbon, and generally much lower concentrations of particulate inorganic carbon.

Total nitrogen concentrations were dominated by nitrate plus nitrite (average about 85 percent) and both exhibited similar patterns with streamflow for each site. Total nitrogen and nitrate plus nitrite concentrations had a positive relation to streamflow at low to moderate streamflows and a negative relation at high streamflows. The studied rivers had nitrate concentrations greater than the drinking-water criteria, or maximum contaminant level, of 10 milligrams per liter in 11 percent of samples, and the draft criteria for aquatic life of 4.9 milligrams per liter (Minnesota proposed limit) was exceeded in 68 percent of samples. Nitrite did not exceed the drinking-water criteria of 1.0 milligrams per liter in any sample. Ammonia was not detected in nearly one-half of all samples, and met the criteria for ammonia, which vary by temperature. Proposed ammonia criteria with more sensitive standards, however, were exceeded in three samples.

Total phosphorus concentrations generally had a positive relation with streamflow, though correlations were site-specific and differed across ranges of streamflow. Total phosphorus concentrations generally increased downstream among Mississippi River tributaries, but were greatest and most variable in southwestern Iowa rivers. Compared to the Wisconsin criteria for phosphorus in large rivers, the studied rivers exceeded the 0.1 milligram per liter standard in 92 percent of samples. Orthophosphate concentrations were flat to slightly positive with relation to streamflow, except the strongly negative orthophosphate-streamflow relation observed in the Boyer River, which also had the greatest concentrations overall.

Suspended-sediment concentrations ranged five orders of magnitude and were positively related to turbidity and streamflow. The greatest concentrations were in the southwestern Iowa basins, which contain extensive loess and steep hills. The positive relation between suspended-sediment concentrations

and turbidity is distinct, but less defined for low values, such that the site-specific relation can be weak [coefficient of determination (R^2) = 0.37] in rivers with typically low suspended-sediment concentrations and turbidity.

Algal pigments and pesticides exhibited strong seasonal patterns. Late-summer algal blooms were evident with peaks in concentrations for chlorophyll-a and pheophytin-a. Spring concentrations and detections peak for pesticides were coincident with application times. Atrazine, metolachlor, and the atrazine breakdown product 2-Chloro-4-isopropylamino-6-amino-s-triazine were detected in all samples. Other commonly detected herbicides included acetochlor, prometon, simazine, alachlor, and metribuzin. Insecticides were less commonly detected, with chlorpyrifos and fipronil detected in 9 percent of samples.

Stream loads are presented for chloride, sulfate, silica, total nitrogen, nitrate plus nitrite, total phosphorus, orthophosphate, and suspended sediment. Because constituent loads in streams are largely determined by streamflow, loads were greatest for all constituents in larger basins and during periods of increased streamflow. For most constituents, the Des Moines River and Iowa River, the largest basins in the study, had the greatest loads of the studied rivers. Constituent yields also were positively related to streamflow, but yields revealed additional spatial patterns in ion, nutrient, and suspended-sediment transport. Chloride yields were greater in the eastern Iowa tributaries to the Mississippi River than in the western Iowa tributaries to the Missouri River. Sulfate yields were greatest in the Big Sioux, Little Sioux, and Des Moines Rivers. Silica yields were lowest in the Big Sioux River. Total nitrogen and nitrate yields were low in the Big Sioux River and greater in the northeastern rivers. Total phosphorus yields were greatest in the Boyer and Nishnabotna Rivers. Orthophosphate yields were greatest in the Boyer River, except in 2008, when the Maquoketa River produced the greatest yield. Suspended-sediment yields were greatest from the Boyer and Nishnabotna Rivers in the southwestern Western Loess Hills region, whereas the Big Sioux and Wapsipinicon Rivers produced the lowest suspended-sediment yields.

Loads presented in this report corroborate previously reported loads for nitrate, orthophosphate, and total phosphorus for the Iowa River at Wapello and suspended sediment for the Skunk River at Augusta. Ranges of predictive errors at the 95-percent confidence limit overlapped for all four instances of comparison. The two instances where predictive errors were presented for both reported loads, the Iowa River at Wapello nitrate plus nitrate and total phosphorus, confidence limits presented in this report were narrower than in previous estimates. This increased accuracy demonstrates the usefulness of the additional streamflow variability terms used in the models. Of the 80 individual site/constituent models, 44 models included streamflow variability terms or streamflow anomalies to improve load estimates. Overall, predictive errors for suspended-sediment loads were greater than most other constituents, with average standard errors of prediction of 21 percent for suspended-sediment loads. Of the other constituents, orthophosphate and total phosphorus were the only others with average standard errors of prediction above 10 percent.

Nutrient loads presented relative to fertilizer distribution indicates that in wet years, fertilizer use and the proportion of nitrogen and phosphorus in the streams relative to use goes up compared to dry years. Loads relative to use also were smaller for nitrogen and phosphorus in the second one-half of the year. Nitrogen loads relative to fertilizer use for January to June (including the critical period of spring application and high streamflow) in the wet 2007–2008 years averaged 52 percent. Phosphorus relative loads for the same periods averaged 9.6 percent.

Acknowledgments

The author would like to thank the numerous U.S. Geological Survey field personnel who collected surface-water data throughout this study and Stephen J. Kalkhoff for help in project development and support. Additionally, three reviewers contributed valuable comments to the betterment of this report.

References Cited

Alexander, R.B., Smith, R.A., and Schwarz, G.E., 2000, Effect of stream channel size on the delivery of nitrogen to the Gulf of Mexico: Nature, v. 403, no. 6771, p. 758–761.

Alexander, R.B., Smith, R.A., Schwarz, G.E., Boyer, E.W., Nolan, J.V., and Brakebill, J.W., 2008, Differences in phosphorus and nitrogen delivery to the Gulf of Mexico from the Mississippi River Basin: Environmental Science and Technology, v. 42, no. 3, p. 822–830.

Arar, E.J., and Collins G.B., 1997, U. S. Environmental Protection Agency Method 445.0, In vitro determination of chlorophyll a and pheophytin a in marine and freshwater algae by fluorescence, Revision 1.2: Cincinnati, Ohio, U.S. Environmental Protection Agency, National Exposure Research Laboratory, Office of Research and Development, 22 p.

Aulenbach, B.T., Buxton, H.T., Battaglin, W.A., and Coupe, R.H., 2007, Streamflow and nutrient fluxes of the Mississippi-Atchafalaya River Basin and subbasins for the period of record through 2005: U.S. Geological Survey Open-File Report 2007–1080, accessed August 28, 2009, at http://toxics.usgs.gov/pubs/of-2007-1080/index.html.

Brenton, R.W., and Arnett, T.L., 1993, Methods of analysis by the U.S. Geological Survey National Water Quality Laboratory—Determination of dissolved organic carbon by uv-promoted persulfate oxidation and infrared spectrometry: U.S. Geological Survey Open-File Report 92–480, 12 p.

Buchmiller, R.C., and Eash, D.A., 2010, Floods of May and June 2008 in Iowa: U.S. Geological Survey Open-File Report 2010–1096, 10 p.

Childress, C.J.O., Foreman, W.T., Connor, B.F., and Maloney, T.J., 1999, New reporting procedures based on long-term method detection levels and some considerations for interpretations of water-quality data provided by the U.S. Geological Survey National Water Quality Laboratory: U.S. Geological Survey Open-File Report 1999–193, 19 p.

Cohn, T.A., 2005, Estimating contaminant loads in rivers—An application of adjusted maximum likelihood to type 1 censored data: Water Resources Research, v. 41, W07003. (Also available at http://dx.doi.org/doi:10.1029/2004WR003833.)

Fishman, M.J., ed., 1993, Methods of analysis by the U.S. Geological Survey National Water Quality Laboratory—Determination of inorganic and organic constituents in water and fluvial sediments: U.S. Geological Survey Open-File Report 93–125, 217 p.

Fishman, M.J., and Friedman, L.C., 1989, Methods for determination of inorganic substances in water and fluvial sediments: U.S. Geological Survey Techniques of Water-Resources Investigations, book 5, chap. A1, 545 p.

Guy, H.P., 1969, Laboratory theory and methods of sediment analysis: U.S. Geological Survey Techniques of Water-Resources Investigations, book 5, chap. C1. (Also available at http://pubs.usgs.gov/twri/twri5c1/.)

Helsel, D.R., and Hirsch, R.M., 2002, Statistical methods in water resources: U.S. Geological Survey, Techniques of Water-Resources Investigations, book 4, chap. A3, 522 p (Also available at http://pubs.usgs.gov/twri/twri4a3/.)

Helsel, D.R., 2005, Nondetects and data analysis: Hoboken New Jersey, John Wiley & Sons, 250 p.

Horowitz, A.J., 2003, An evaluation of sediment rating cu for estimating suspended-sediment concentrations for subsequent flux calculations: Hydrological Processes, v p. 3,387–3,409.

Iowa Environmental Protection Commission [567], 2002 Chapter 61 Water quality standards, section 3, Surface quality criteria, Iowa Administrative Code.

Koltun, G.F., Eberle, Michael, Gray, J.R., and Glysson, G.D., 2006, User's manual for the graphical constituent loading analysis system (GCLAS): U.S. Geological Survey Techniques and Methods, book 4, chap. C1, 51 p. (Also available online at *http://pubs.usgs.gov/tm/2006/tm4C1.*)

Monson, Phil, 2010, Aquatic life water quality standards technical support document for nitrate—Triennial water quality standard amendments to Minn. R. chs. 7050 and 7052—DRAFT for external review: St. Paul, Minn., Minnesota Pollution Control Agency, 21 p. (Also available at *http://www.pca.state.mn.us/index.php/view-document. html?gid=14949.*)

Nalley, G.M., Gorman, J.G., Goodrich, R.D., Miller, V.E., and Housel, K.S., 2005a, Water resources data Iowa water year 2004—volume 1, surface water and precipitation: U.S. Geological Survey Water-Data Report IA–04–1, 473 p.

Nalley, G.M., Gorman, J.G., Goodrich, R.D., Littin, G.R., Linhart, S.M., Miller, V.E., and Housel, K.S., 2005b, Water resources data Iowa, water year 2005: U.S. Geological Survey Water-Data Report IA–05–1, 552 p. (Also available at *http://pubs.usgs.gov/wdr/2005/wdr-ia-05-1/.*)

National Oceanic and Atmospheric Administration, National Climate Data Center, 2003, U.S. climate normals 1971–2000, accessed January 29, 2009, at *http://www.ncdc. noaa.gov/oa/climate/normals/usnormals.html.*

O'Dell, J.W., ed., 1993, U. S. Environmental Protection Agency Method 365.1 Determination of phosphorus by semi-automated coloritetry, revision 2.0: Cincinnati, Ohio, U.S. Environmental Protection Agency, Environmental Monitoring Systems Laboratory, Office of Research and Development, 17 p.

Patton, C.J., and Kryskalla, J.R., 2003, Methods of analysis by the U.S. Geological Survey National Water Quality Laboratory—Evaluation of alkaline persulfate digestion as an alternative to Kjeldahl digestion for determination of total and dissolved nitrogen and phosphorus in water: U.S. Geological Survey Water-Resources Investigations Report 03–4174, 33 p.

Runkle, R.L., Crawford, C.G., and Cohn, T.A., 2004, Load estimator (LOADEST)—A FORTRAN program for estimating constituent loads in streams and rivers: U.S. Geological Survey Techniques and Methods, book 4, chap. A5, 75 p.

Stenback, G.A., Crumpton, W.G., Schilling, K.E., and Helmers, M.J., 2011, Rating curve estimation of nutrient loads in Iowa rivers: Journal of Hydrology, v. 396, p.158–169. (Also available at *http://dx.doi.org/doi:10.1016/j. jhydrol.2010.11.006.*)

TIBCO Software Inc., 2008, TIBCO Spotfire S+® 8.1 for Windows® user's guide: Palo Alto, Calif., 582 p. Accessed February 17, 2012, at *http://stn.spotfire.com/stn/UserDoc. aspx?UserDoc=spotfire_client_help%2fintro%2fintro_ introduction.htm&Article=%2fstn%2fDefault.aspx.*

U.S. Census Bureau, Population estimates program, accessed January 27, 2011, at *http://www.census.gov/popest/ estimates.php.*

U.S. Department of Agriculture, 2009, 2007, Census of agriculture, United States summary and state data, volume 1: Geographic Area Series, part 51, 639 p.

U.S. Environmental Protection Agency, 1999, Update of ambient water quality criteria for ammonia: EPA–822–R–99–014, National Technical Information Service, Springfield, Va.

U.S. Environmental Protection Agency, 2002, Section 141.61 Maximum contaminant levels for organic contaminants, Code of Federal Regulations Title 40 Protection of Environment, U.S. Government Printing Office, Washington, D.C.

U.S. Environmental Protection Agency, 2009a, Draft 2009 update of ambient water quality criteria for ammonia— Freshwater: EPA–822–D–09–001, National Technical Information Service, Springfield, Va.

U.S. Environmental Protection Agency, 2009b, Ecoregion maps and GIS Resources, accessed November 9, 2009, at *http://www.epa.gov/wed/pages/ecoregions.htm.*

U.S. Geological Survey, 2007–2009, Water-resources data for the United States, accessed May 1, 2009, at *http://wdr.water. usgs.gov/.*

U.S. Geological Survey, variously dated, National field manual for the collection of water-quality data: U.S. Geological Survey Techniques of Water-Resources Investigations, book 9, chaps. A1–A9. (Also available at *http://pubs.water. usgs.gov/twri9A.*)

U.S. Geological Survey, 2009, National Water Information System (NWISWeb): U.S. Geological Survey database, accessed May 1, 2009, at *http://waterdata.usgs.gov/nwis/.*

Vecchia, Aldo, 2003, Relation between climate variability and stream water quality in the continental United States: Hydrological Science and Technology, v. 19, no.1, p. 77–98.

Vitousek, P.M., Aber, J.D., Howarth, R.W., Likens, G.E., Matson, P.A., Schindler, D.W., Schlesinger, W.H., and Tilman, D.G., 1997, Human alteration of the global nitrogen cycle—sources and consequences: Ecological Applications, v. 7, p. 737–750.

Wang, Ping, and Linker, L.C., 2008, Improvement of regression simulation in fluvial sediment loads: Journal of Hydraulic Engineering, v. 134, no. 10, p. 1,527–1,531.

Wetzel, R.G., 1983, Limnology, 2nd ed., Philadelphia, Saunders College Publishing, 767 p.

Wisconsin Department of Natural Resources, 2010, Water quality standards for Wisconsin surface waters: chapter NR 102.06, Phosphorus, Register, November, 2010, no. 659. (Also available at *https://docs.legis.wisconsin.gov/code/admin_code/nr/102/I/06.*)

Zimmerman, C.F., Keefe, C.W, and Bashe, Jerry, 1997, U.S. Environmental Protection Agency Method 440.0, Determination of carbon and nitrogen in sediments and particulates of estuarine/coastal waters using elemental analysis, revision 1.4: Cincinnati, Ohio, U.S. Environmental Protection Agency, National Exposure Research Laboratory, Office of Research and Development, 10 p.

www.ingramcontent.com/pod-product-compliance
Lightning Source LLC
Chambersburg PA
CBHW081607170526
45166CB00009B/2858